Saga of the Steam Plough

Saga of
the Steam Plough

HAROLD BONNETT

DAVID & CHARLES
NEWTON ABBOT

ISBN 0 7153 5742 5

First published by George Allen & Unwin, 1965

New impression published by David & Charles, 1972

Set in 11 pt Plantin
and printed in Belgium
by Imprimerie Jos. Adam
for David & Charles (Publishers) Limited
South Devon House Newton Abbot Devon

PREFACE

Of all man's inventions, the steam engine has possibly more appeal than any, not excepting the sailing ship. With both railway locomotive and traction engine there is motionwork and clouds of smoke, rhythm to feel, and a soul elevating sound of exhaust. The type of traction engine known as the ploughing engine, with which the author has dealt so effectively, was the very quintessence of steam engine magnetism.

When cable ploughing engines were at work, there was the stately exhaust beat of the old 'Singles', which could rise into a deep bellowing with spark threaded black smoke rising into the evening air; or there would be the softer panting of the lower geared 'Compounds' with their whizzing flywheels. Sometimes, a pair of engines might be found, one silhouetted on the horizon and the other working down in an oak shaded hollow or backed by a high hawthorn hedge, effectively camouflaged, and only the tell-tale grey-flecked smoke rising above the tree tops, as it started up in its turn, would give it away.

Most engines evoked endearing emotions, and drivers often merged their personalities with them. And what men! The best born to it and brought up on the ploughing sets, starting as cook boys. But there were also journeymen who had learnt their steam craft later, or in other places. Dragmen, recruited perhaps from farms after having been with the water cart; men off the roads; men who had been to sea; or fairground engine drivers; steam roller men; all sorts, but welded into a team when under a good foreman. I knew, mostly by nicknames, a few around Bedfordshire and Northamptonshire: Tommy; Darner; Tark; Poggs; Nip; Steam Ted; Tiger; Lincoln Bill; Dawson; Brownie and Bozzy. All now gone, yet surely immortal, because if there is a Heaven there will be English fields of long ago in it, and a little steam ploughing done here and there.

E. E. KIMBELL
Boughton
Northampton

CONTENTS

Preface   *page* 7

   I *The Years of Ideas*   17

  II *The First Steam Plough*   20

 III *The Middle Period of Invention, 1834–54*   24

 IV *John Fowler*   34

  V *Oxman and Horseman to Steam Ploughman*   43

 VI *The Royal Shows*   47

VII *Other Notable Pioneers*   50

VIII *Iron Horse v. Hided Horse*   57

 IX *Steam Plough Firms and the Engines they made, 1860–1900*   65

  X *Steam and the Steam Plough*   83

 XI *Direct Tractionists at Work*   88

XII *The Steam Plough in North America*   95

XIII *Steam Power Implements*   100

XIV *Steam Cultivation Contractors*   108

 XV *The Traction Engine challenges the Motor Tractor*   120

XVI *Steam Ploughing as a Pastime*   126

XVII *The Steam Plough Owners organize for War*   140

XVIII *British built Steam Ploughs at work around the World*   148 *page*

XIX   *Some Final Types of Lightweight Steam Engines for Ploughing*   154

XX   *Steam Ploughing in Germany*   159

XXI   *The Last Years of Commercial Activity with Steam Ploughs in Great Britain*   165

XXII   *The Steam Plough in World War II*   180

XXIII   *Boiler Explosions*   183

XXIV   *The Last Chapter of Active Steam on Land*   186

XXV   *What was Wrong with Steam Ploughing?*   194

XXVI   *Preservation Days*   198

Appendix: Principal Ideas and Achievements associated with Steam Cultivation   200

Index   205

ILLUSTRATIONS

1. Heathcoat and Parkes' steam plough, 1833
   Smith's steam cultivator   *facing page* 32

2. Fowler's patent double cylinder plough engine
   Fowler's double engine set of ploughing tackle at work, 1880
   facing page 33

3. 1856 Burrell Engine with Boydell 'Endless Railway' wheels
   Savory's Drum-round-the-Boiler Ploughing Engine of 1861
   facing page 48

4. Lord Willougby's Steam Plough of 1850   *facing page* 49

5. McLaren 20 n.h.p. compound plough engine
   View of Eddison's yard at Dorchester, about 1885
   facing page 64

6. Aveling and Porter single engine steam ploughing tackle
   1907 Aveling and Porter compound type plough engine
   1911 Aveling and Porter 30 n.h.p. compound steam plough
   engine   *facing page* 65

7. Howard 'Farmer's Friend' single engine ploughing tackle
   Burrell 16 n.h.p. single cylinder plough engine
   facing page 80

8. Darby steam digger working at Peldon, Essex, 1904.
   Wallis and Steevens steam tractor with plough bodies
   attached either end   *facing page* 81

9. Fowler 14 n.h.p. single cylinder plough engine
   'Oxford' 12 n.h.p. compound steam plough engine
   facing page 96

10. Fowler plough engine on beach at Gallipoli, 1915
    Fowler plough being used for cutting trenches
    facing page 97

11. U.S.A. steam plough engine for direct traction work
    German Borsig steam plough with vertical engine and rope
    drum at rear   *facing page* 112

12. Lang's system of wire rope construction
    A fine pair of 1918 Fowler B.B. compounds   *facing page* 113

13. Fowler 'Superba' superheated steam plough engine
    Mann lightweight agricultural steam tractor   *facing page* 128

14. 5 n.h.p. Garrett light three-shaft single cylinder traction engine
    14 n.h.p. single cylinder Fowler after it had blown up
    facing page 129

15. 'Old time' demonstration of steam ploughing
    Fowler 'AA' No. 14726 at work cultivating   *facing page* 144

16. Lightweight cable steam plough built in 1927
    Crankshaft and gear wheel pinion   *facing page* 145

17. A Fowler takes time off to pull up some trees
    Sale of John Patten's steam ploughing equipment
    facing page 160

18. 1918 Fowler BB No. 15226 'Tiny Tim'
    Driver's view of Fowler six-furrow anti-balance plough
    facing page 161

AUTHOR'S ACKNOWLEDGEMENTS

The author wishes to acknowledge all the help he has received. Many of the persons or firms who have so willingly provided information are, of course, mentioned specially in the book. The author is particularly appreciative of the constructive criticisms made by the Rev. R. C. Stebbing, Mr. E. E. Kimbell and Mr. Douglas Bomford, after they had read the original manuscript. Others to whom the author is also grateful are:

Mr. G. V. S. Andrews, Sleaford, Lincs.
The Earl of Ancaster, Drummond Castle, Perthshire
Mr. A. C. Durrant, Kensal Rise, London
The late Mr. T. Boughton, Amersham, Bucks.
Dawson & Sons, Agricultural Engineers, Frampton, Boston
Mr. A. James, Hewelsfield, Glos.
Mr. E. J. King, Strood House, Peldon, Essex
Lincolnshire Archives Committee, Lincoln
Major R. D. Marshall, Gainsborough, Lincs.
Mr. Alf Pepper, Fowler Veterans' Association, Leeds
Road Locomotive Society (For access to Portfolios)
The United Africa Company Ltd.
Librarians at: British Museum, Ministry of Agriculture,
     Guildhall in London, The War Office,
     Science Museum, Patent Office, and
      U.S.A. Information Service, London.

My wife Hilda has also made a valuable contribution by reading and improving the final manuscript.

3, Ridge Side,
Haw Lane,
Bledlow Ridge,
Nr. High Wycombe,
Bucks.

Book of Farm Implements & Machinery, 1858–Slight and
    R. Burn-Scott
Cassier's Magazine
A Century of Traction Engines—W. J. Hughes
Chambers's Encyclopedia
The Cornish Giant—L. T. C. Rolt
Diesel Railway Traction
Down on the Farm—S. Holbrook (U.S.A.)
Engineer, The
Engineers' & Mechanics' Encyclopedia, The
Farmers' Tools, The—G. E. Fussell
Farm Power—Moses and Frost (U.S.A.)
God Speed the Plow—Clark C. Spence (U.S.A.)
Implement & Machinery Review
Journals of the Royal Agricultural Society of England
Mechanics Magazine, The
Model Engineer
Minutes of Steam Cultivation Development Association
    Meetings
Steam Engine Builders of Norfolk—Ronald H. Clark
Die Technik in der Landwirtschaft (Germany)
Unpublished Manuscript by Theo. Davis (Part History of
    John Fowler & Co. Ltd.)

I

The Years of Ideas

Patent No. 6, granted in 1619 by James I to his Page of the Bedchamber David Ramsey, and to Thomas Wildgoose, is the probable beginning to the story of steam ploughing. Part of this patent reads:

'Who have by theire industrye, and att theire greate paynes, costs, charges and expences, devised, found out, and brought to pfection, divers Newe, Apte, or Compendious Formes or Kind of Engines or Instruments to Ploughe Ground, without Horse or Oxen. . . . They have the sole privilege of making and using said engines. . . . Any infringement will be defaced, thrown down, destroyed and broken. . . .'

Unfortunately we do not know for sure whether it was actually a steam engine which these patentees had in mind. It is a fact, however, that the Marquis of Worcester, who was born in 1601, had by 1663 devised and constructed a crude but workable steam pumping engine. It is not so unlikely, therefore, that the patentees had at least heard court gossip of the investigations already afoot to utilize steam power. The first really practical steam engines were, of course, built by Thomas Newcomen and Captain Thomas Savery, during the early years of the following century.

In 1767, Francis Moore, a linen draper in Cheapside, London, came forward with an idea described as, 'A Fire Engine for Ploughing', although the patent which he obtained does not give us any helpful mechanical detail. As it was common practice right down into the nineteenth century to describe steam engines as 'Fire engines' (because they needed a fire to raise steam) it is fairly safe to assume Moore was thinking of a steam engine. He seems to have sold most of his horses, and advised his farmer friends to do the same, believing his proposed steam engine would rapidly do away with horses for ploughing.

James Watt, the talented Scottish steam pioneer, well known

from the story of how as a boy he watched the force of the steam making the lid of his mother's kettle bob up and down, also patented in 1770 a steam carriage which he considered would be able to pull agricultural implements. He believed in low pressure steam, however, holding the view that it would never be possible to construct boilers to withstand pressure above, say, 10 lb. per square inch. Watt's chief claim to fame rests upon his invention of the double-acting beam engine, one in which steam was for the first time admitted alternately to either side of the piston. Previously, the Savery and Newcomen engines had been of the type where steam was admitted into the lower half of the cylinder, and immediately condensed by a squirt of cold water to create a partial vacuum. It was the atmospheric pressure acting on the upper and open side of the piston which forced it down and so provided the real power stroke of these engines.

Five years before Waterloo, in 1810, patent No. 3309 was granted with the customary form of salutation, 'To all to whom these presents shall come, I, Major Pratt of Spencer Street, in the Parish of Saint George in the East, in the County of Middlesex, Farmer, send greeting. . . .' The patent then goes on to give a detailed description of several kinds of farm implements; each idea is splendidly illustrated by etched drawings. Even more praise-worthy is the original thought behind the ideas which include an endless running chain to which most farm implements could be attached, a rotary cultivator, and a plough with fixed shares. Finally, there is a most interesting arrangement of two plough-shares and mouldboards facing in opposite directions, but pivoted centrally to provide a balance point on which to tilt the frame for two-way working without turning. This seems to be the first attempt to get over the problem of how to work a plough on the cable system and avoid having to turn it round at the end of the field. One rotary cultivator is shown powered by four windmill sails turning on a horizontal plane, helicopter fashion, over the implement. A strong gale would have been needed to work this machine. This patent relates to implements only, and in Pratt's own words their haulage is dismissed as '. . . being connected with wind, steam or any other such like mechanical power, which any ingenious workman will know how to do. . . .' A bit airy fairy about power perhaps, but we could hardly expect Pratt to invent implements and engines at the same time.

Into the early realm of steam there now appeared, at Camborne in West Cornwall, a remarkable personality in the form of Richard Trevithick. He was a burly young man bubbling over with enthusiasm; he turned out to be a talented, versatile and successful steam engineer with brilliant new notions. Unlike Watt, he believed in high pressure steam, realizing from his experience with the cumbersome beam engines pumping water from the Cornish tin mines, that if sufficient power in relation to weight were to be obtained, the only answer was steam at pressure of at least 100 lb. per square inch. Unfortunately, although in 1801 he built the second road locomotive in Great Britain (the first was by W. Murdock in 1784 in model form), and in 1804 the first tramway locomotive in the world, he did not take his steam plough ideas beyond designing in 1813, a cultivating machine intended for haulage behind a steam engine. While Major Pratt had arranged the rotating tines of his implements to move on horizontal planes, Trevithick stipulated a large and fearsome-looking rear wheel mounted vertically. Indirectly, of course, Trevithick made a rich contribution to future steam plough development, by constructing boilers which held high pressure steam safely.

Other inventors of this period are listed in the appendix at the end of the book; but doubtless there are many who are not mentioned. Some are now quite unknown to us, but we may be sure that in the great country houses of the aristocracy, in the small ironfounders' sheds, or in the shops of village blacksmiths much thought was given and trial work carried out by unrecorded pioneers. All that had been done was at last beginning to bear fruit. The age of accomplishment was in sight.

II

The First Steam Plough

In the 1830's British agriculture was in a relatively good position. While labourers' wages were low, many large farmers as well as the landed gentry were reasonably prosperous. The Industrial Revolution was gathering momentum across the Midlands and the North, but agriculture was still the principal and most important industry. Macamadization of the main roads had begun; and over the new hard surfaces of small stones packed closely together, rattled a service of fast stage and mail coaches. These conveyances not only carried the new steam engineers up and down the country, but also the newspapers and periodicals of the times containing articles on the experimental work mentioned in the previous chapter. Information now began to flow more freely and widely, and there was sufficient money available to oil the wheels of ideas. Over the fields of quiet, well-wooded and mostly unspoilt England was soon to blow the smoke of the first steam plough.

If we leave aside the infant railways with their 'Puffing Billy' locomotives, the steam engines so far constructed were, apart from such isolated examples as Trevithick's road engine of 1801, or the lightweight designs for steam road carriages, chiefly of the stationary type. These heavy, cumbersome and low-powered engines were not good enough for the inventors who wanted a separate, light and mobile engine which could be hitched to the front of the plough in place of the team of horses or oxen. There was no engine really suitable for them. It is not surprising, therefore, to find that others had already given some thought to a system by which the heavy power plant itself stood still while it hauled the implement. A rope, chain or other flexible band was to be used to pull the mobile implement between the standing engine, and a pulley anchored on the opposite side of the field.

A machine of this description was introduced by two remarkable men, John Heathcoat, M.P. for Tiverton in Devon, a lace manu-

20

facturer with at least 23 textile trade patents in his name, and Mr Josiah Parkes, an expert drainage engineer. Working in partnership, they succeeded in conceiving, constructing and demonstrating the first practical steam plough. Their patent No. 6267 was taken out in 1832, and was followed by the appearance of the machine the next year. In 1835, extensive trials were carried out on boggy ground at Red Moss near Bolton-le-Moors, Lancashire. This tackle had been designed expressly for ploughing bog land, as Parkes' was chiefly interested in drainage and land reclamation. Their steam plough was certainly an oddity.

The power unit comprised a low level platform carrying a two cylinder steam engine coupled to two winding drums. The whole assembly was slung like a huge litter between two end axles, upon each of which were mounted six wheels eight feet in diameter. Over the narrow rims of these tall wheels, ran the engine-driven, endless plank track by which the machine, which weighed thirty tons, was propelled a few feet ahead, after each pull at the plough. Friction rollers in the form of small wheels were interposed on the underside between frame and moving track, which at the top was carried rooflike over this mobile powerhouse. Although it is of great interest to find an endless track in use at such an early period, in 1821 John Richard Barry of the Minories, London, had devised a system of propulsion using an endless chain to couple eight two-wheeled axles, moving in the manner of an endless track. Heathcoat and Parkes had improved on Barry's invention, but in any event the caterpillar type of propulsion was an innovation of great promise which did not gain full recognition until eighty years later, during the First World War, when it was used with sensational success on British tanks.

The plough was perhaps even more odd than the engine. It had cutting knives to snip and slash through the matted mass of plant fibres and roots found in all bog land. One submerged and single-bladed knife reciprocated horizontally across the front of the ploughshare. A second scissor-action knife made a vertical cut ahead of the coulter, just like a tailor cutting thick cloth lying on a table. Both these knives were driven by a spiked wheel running on the unploughed part of the bog. It was a single furrow plough with a pair of ploughshares and mouldboards placed end to end, but facing in opposite directions. At the end of the furrow the two ploughmen, and the man who helped with the winding gear, manhandled the $12\frac{1}{2}$-cwt plough into position for the return run. In spite of the long handles provided, it must have been a shocking job to swing the implement over into the next furrow line.

Two flexible flat iron plate bands $2\frac{1}{8}$ inches wide and $\frac{1}{16}$ inch thick, riveted together in short sections, were used as a form of rope to haul the plough backwards and forwards between the engine and a movable auxiliary carriage placed directly opposite, at about 220 yards distance. Large winding drums on the engine platform wound these bands in and out. When the fields were wide enough, two ploughs were used, one on either side of the machine. The main trouble appears to have been the continual breakage of the flexible bands.

In November 1835, Henry Hanley, of Culverthorpe Hall, Lincs., addressing the Grantham Agricultural Society said he had recently seen this steam plough doing very good work in Lancashire. It had ploughed an acre 9 inches deep in 1 hour 50 minutes.

At the Highland and Agricultural Society's Show at Dumfries in 1837 opportunity was taken to stage a further demonstration in the hope of winning the £500 prize for a successful method of ploughing by steam. The tackle was transported from Lancashire mainly by sea and set up on Lochar Moss, where it gave a reasonable performance on the first day, ploughing eight acres of bog in twelve hours. They say crowds of country folk walked miles to watch spellbound this extraordinary exhibition. In places the moss was so soft and wet that on several occasions the two ploughmen were submerged up to their waists. At the end of the day, the opinion of the judges was that the machine might well be adapted for the cultivation of ordinary land.

Nobody seems to know what became of this first steam plough. The story goes that during the first night after the trials, the whole contraption toppled over and sank below the surface of the swampy morass. It was never seen again, and although there is no confirmation of this story, there may be some truth in it. It was one of the recommended preliminaries before starting work that two deep ditches spaced twenty-two feet apart should be dug, leaving a central strip upon which the machine might travel as it edged its way slowly ahead during ploughing. It was Parkes' ultimate intention to drain the Moss and use the consolidated strips as permanent farm roadways. It is just possible that one side of the springy and spongy strip upon which the machine stood, slipped into the water, and if this happened the whole contraption could have pitched sideways into the bog. Interested people still go and look over the locality, in the hope they will be able to solve this great mystery.

Seven men and a boy were required to operate the tackle and, in all, Heathcoat spent no less than £12,000 on its construction and demonstration, a very large sum for those days. Although the project was bold and imaginative, and reasonably successful, its investors reaped only financial loss and a fair measure of disappointment. They had, however, done some good pioneering work for which we ought to be grateful. Things were now moving in the right direction—they had proved that steam could be made to pull a plough.

III

*The Middle Period of Invention
1834–54*

This period, following as it did the appearance of the first steam plough, was one of the most prolific for new inventions. The idea of steam cultivation had caught on. It is significant, too, that whereas previously machines were usually left in the proposal or outline drawing stage, construction soon became the rule rather than the exception.

Although there is no record it was ever built, the steam carriage designed by John Upton in 1837 is of particular interest. It was to have a water tube boiler and an engine functioning on the rotary action principle instead of the usual reciprocating piston arrangement. Four sliding vanes, centred on an eccentric shaft passing through a drum type cylinder, were intended to give a continuous rotary motion in a similar manner to that obtained from a steam turbine. This lightweight vehicle was described by its inventor as suitable for ploughing, harrowing and other agricultural purposes.

In 1839 a Colonial inventor, Alexander MacRae of Demarara, British Guiana, giving his address as the London Coffee House, Ludgate Hill, London, took out a British patent for a single engine cable plough system. The flat, low-lying coastal strips of Demarara were intersected with drainage dykes in the form of parallel navigable canals; and the tackle was expressly designed for cultivating the smallish flat fields bounded by these waterways. The engine with its winding drums was to be in a large punt, with an endless chain carried out to a free-running pulley fixed in a second punt floating in a parallel canal about 120 yards away. The proposed four-wheel plough had three furrows for working in either direction, with each ploughshare and mouldboard hinged separately and lifted by its own hand lever similar to an American gang plough, i.e. gang or cluster of independently slung shares and

24

mouldboards. So far as is known, however, no actual construction was undertaken.

Henry Pinkus, who in 1839 styled himself as a Gentleman of Old Slaughter's Coffee House, St Martins Lane, is also a name to remember for an unusual inventive bent on cultivation. He proposed a two-way plough powered by vacuum supplied through underground pipes from a central steam power house. The reel of $\frac{1}{2}$ inch internal diameter flexible hose on the plough would be connected to the main vacuum standpipes, uncoiling or recoiling as the implement traversed to and fro over the fields. In 1840 Pinkus estimated there were a million horses on English farms worth £30,000,000 and that already large sums of money had been spent attempting to adapt steam locomotive engines to displace them. Owing to the excessive weight of the engines, however, all such schemes had failed but he hoped his might prove successful, if for no other reason than the humane one that slave labour abroad might be abolished. A model of his apparatus was exhibited at the Colosseum in Regent's Park, but again there is no proof of actual construction. Incidentally, Pinkus also experimented unsuccessfully with vacuum power for the Birmingham, Bristol and Thames Junction Railway then under construction near Kensington. Although the theme of this history is primarily about steam, perhaps it ought to be mentioned that in 1839 Pinkus drew up a patent specification for an internal combustion-engined farm tractor. Ambitious again, he visualized small dispersed gas works generating carburetted gas, which like the vacuum, was to be distributed through underground pipes. Ignition was to be effected by means of a permanent flame burning alongside a four-way valve on the engine cylinder head, something after the manner of firing an old breach-loading cannon through the touchhole.

As later successful steam ploughing engines were mostly of the traction engine type, the exhibition by Ransomes, Sims and Head, Ipswich, of a chain-driven self-propelled engine-cum-thresher with a vertical boiler at the 1842 Royal Show at Bristol must be mentioned. This was the very first traction engine to be exhibited at any agricultural show, and was certainly one of the earliest of all self-moving heavy road engines.

Another Colonial from Demarara with an urge to improve agricultural machinery, was John Tulloh Osborn who came to

London and patented his ideas in 1846. In his specification he mentions steam, water, air or vacuum as possible prime movers for implements, but he achieved fame in concentrating on the introduction of what was most likely the first two-engine steam ploughing tackle. While the actual year of introduction is uncertain, his first engines were designed with two vertical side winding drums upon which long chains were coiled, so that the ploughs attached to them could be drawn across the fields from one engine to the other. The two cylinders were located, as on railway locomotives, under the engine smokeboxes, where they were conveniently situated to drive the drums. Although not self-moving in the ordinary sense the engines pulled themselves forward a few feet at a time by winding in a third chain anchored ahead; this kept them abreast of the ploughing in progress at right-angles to their own courses alongside opposite hedgerows. To prevent the wheels from sinking into wet ground, each engine moved and stood upon wooden rafts made in short sections.

Two ploughs were used at the same time, each moving in opposite directions. Their slack chains were carried outwards on the end of a rigger, to clear the other plough, and leaving the chain in position ahead ready for the return journey. This meant that one drum on each engine was hauling in a plough, while the other drum was paying out slack chain to the other plough as it receded towards the opposite engine. In order to make it easier to swing the two-way plough (one which was worked both ways without turning) over into the new breadth of land at the ends, portable ramps were laid alongside each engine at the finish of the run. This ramp lifted the ploughshares out of the ground and so gave free movement. At this early date, it was much to Osborn's credit that he introduced such workable tackle. The idea of having two engines and using both at once with separate ploughs was especially significant.

Osborn was fortunate to find a keen and generous patron in George the Marquis of Tweeddale, a member of the Scottish aristocracy said to be proficient in mechanics and an agricultural inventor to boot. Besides being tall, strong and famous as a boxer, the Marquis was also an accomplished horseman popularly known as the 'Prince of the Heavy Bays'. On one occasion he drove at one sitting the Mail Coach from London to Haddington, a distance of 368 miles. Based on Osborn's designs, a two-engine ploughing set was built and given extensive trials at Yester, the family seat in East Lothian. Even though it is not certain the first tackle was strictly in accordance with Osborn's original drawings, it is very likely that with a few modifications suggested by the Marquis, it was substantially the same. Extensive experimental steam plough work continued at Yester over many years, with various types of engines and implements. I think it can truly be said this sporting, inventive and colourful Scottish laird was the great pioneer of steam ploughing in Scotland. He died in 1876 aged eighty-nine.

Side by side with the development of the plain plough, the inventors of rotary cultivators pressed on in an endeavour to secure the lead. Men such as Atkins, Paul and Sir John Scott Lillie either made proposals or took out patents. James Usher, a gentleman brewer of Edinburgh had built in 1849 by Messrs Slight, agricultural engineers of Leith, a direct traction type rotary digger. Really it was a portable type steam engine, mounted on the same four-wheel frame as the digging gear which was fixed behind. Toothed gearing transmitted the engine power to the road wheels, as well as to the digger cutters, which rotated in the same direction as the main wheels, so that the digging action also helped to push the engine forward. Rated at 10 h.p. and weighing $6\frac{1}{2}$ tons, this engine-cum-digger could cultivate 7 acres a day while travelling at a speed of 3 miles per hour. Even though this machine was not an unqualified success, it was capable of doing a satisfactory job of work and encouraged Usher to continue his experiments in Scotland with improved and similar diggers over the next five years or so with reasonable satisfaction.

The wind next blew in favour of the plough hauled by indirect means. Peter Robert, Lord Willoughby d'Eresby, another steam-minded man of rank and means, contributed greatly in personal interest and money, to the development of single and double engine sets. A close friend of Daniel (later Sir Daniel) Gooch, the celebrated locomotive engineer of the Great Western Railway, 1837–64, who designed some of the first practical engines for that line, Lord Willoughby was, by this connection, able to arrange for what appears the only steam ploughing engine ever to be built in a railway locomotive works. From old G.W.R. records, now in the possession of the British Transport Commission's Archives Library, I am able to quote the following extract from a letter to the company's secretary which began about materials for G.W.R. tank engines:

Swindon,
November 30th 1849.

Charles A. Saunders,
Paddington.

My dear Sir,

. . . .

I shall be glad if the Directors will sanction my building a small engine at Swindon for Lord Willoughby d'Eresby, the cost exclusive of wheels and which he can supply us with on better terms than we can make them to be £305.

Yours very truly,
Daniel Gooch.

When completed, this engine was named 'California' and sent down to Grimsthorpe Castle, west of Bourne, Lincolnshire where trials were subsequently undertaken on his Lordship's extensive estate. After it was delivered in May 1850, Lord Willoughby and Daniel Gooch went there together to see it tested at a saw bench and a corn mill. Unfortunately, it was not quite as powerful as they expected; it was thought this was due to the flywheel being too small, and a temporary wooden one was made on the estate, until a new larger one could be manufactured in cast iron. In its favour, however, it must be said that G. G. Scott, the agent at Grimsthorpe when writing to Lewis Kennedy, the steward of all Lord Willoughby's estates said, 'It (California) is the most beautiful piece of workmanship I have ever seen'. Mr Scott's correspondence contains a wealth of the most useful and revealing information about the five years of engine trials at Grimsthorpe, from which it emerges there were at least two other engines, the 'Juba' and the 'Oasis', and although the handwriting is not now quite legible, it rather looks as if 'Juba' is described as a 'Hornsby' engine (R. Hornsby & Sons, Grantham). All the engines were used for ploughing at some time or other, whether as single units or in two-engine sets. 'California' was rated at 26 h.p. and although she weighed only $3\frac{1}{2}$ tons had in addition a coke and water tender hooked on behind. It was not self-moving, although with his design of two cylinders over the smokebox, interposed valve link motion and crankshaft above the firebox, Gooch was obviously a visionary in regard to future traction engine practice. Even the bevel drive to the underslung chain drums was a portent of the arrangements incorporated almost universally in later years for drum driving gear. When ploughing this engine also stood upon wooden rafts, known at Grimsthorpe as a portable railway, only in this instance there were channels or guideways to hold the engine wheels on a straight course.

The Estate carpenters appear to have been kept busy making and repairing these portable railways, the ploughs and other wooden parts for the tackle. So much so in fact that Scott complained to Kennedy in January 1851, 'The Carpenters have been doing nothing but working at things for this steam ploughing . . . & until his Lordship has been down and made a trial, I believe we shall get no work done. We might all the time have been making door frames . . . etc., for repairing cottages, but we are getting fresh orders every day from his Lordship.'

When he came down to Grimsthorpe in May, Lord Willoughby was not particularly pleased with the way things were going. But when Mr J. Finley who had accompanied him complained of the great number of men required at the ploughing and also the great expense, his Lordship soon made the sharp answer 'Although all the World find fault with it, I will not be prevented from carrying out my plans'. This I think was a typical piece of British bulldog tenacity on the part of the noble Lincolnshire landowner, and helped much to establish eventually the custom of ploughing by steam.

The plough chains used by Lord Willoughby obtained their driving grip by being wound one and a half turns round each engine drum, from which they extended 180 yards to a free-running pulley wheel fixed on a four-wheeled anchor cart placed on the opposite headland. Proper tension was maintained by means of a screw adjuster attached to the pulley; but unless constant attention was given, one can well imagine the chains slackening and slipping on the drums. Indeed, it seems as if the clumsy chain haulage was the greatest source of trouble with this Grimsthorpe machinery.

By the clever arrangement of having the ploughshares in the ground and ploughing only on the inward trip towards the engine, the strain on the anchor carts never exceeded the comparatively easy load of pulling back the raised or 'Empty' ploughs as they returned, running on two small hinged wheels. With single engine working, both drums could be used to operate two ploughs, one on either side of the engine; and each moving to and fro between it and the two flanking anchor carts. When two-engine working was in progress, only one drum on each was used; because there is mention of the time wasted while the 'Empty' ploughs were pulled back. Two engines with a two-furrow plough were taken from Grimsthorpe to the Great Exhibition of 1851 in Hyde Park.

After the prime object of proving the mechanical practicability of these early steam tackles, the next important and crucial requirement was that, if they were to succeed, the work must be done as cheaply as by horses or oxen. Some interesting figures on double-engine ploughing costs were prepared in April 1853 at the personal request of Lord Willoughby:

	£	s.	d.
1st engine driver @ 3/4d. per day		3	4
2nd engine driver @ 2/10d. per day		2	10
4 labourers @ 1/8d. per day		6	8
1 boy @ 8d. per day			8
Horse and water cart—2 boys per day		5	6
Coke for two engines per day		10	0
Total =	1	9	0

As four acres were turned over daily, the cost was 7/3d. per acre for the ploughing which farmer observers said was quite as well done as with horses, but no cheaper. Had charges for interest on capital, repairs or removal by horses from field to field been included, the real costs would, of course, have been a bit higher. At ten shillings a week labourers' wages were pitifully low, but even so the men afforded some pleasure in life, for we find the agent writing an official complaint about the steam plough gang going to an all-night dance prior to a trial when Lord Willoughby d'Eresby intended to be present. Whether they were late, or tired out, or whether any man was sacked is not recorded, but they appear to have been a typical bunch of happy-go-lucky steam ploughmen.

An application was made by a certain John Johnson on December 23, 1854, for the tenancy of the Steam Plough Inn at Little Bytham, a village nearby. Unfortunately, my enquiries trying to identify this pub with any of the present or recently closed 'houses' were unsuccessful. I have not heard of any other inn so named; and it is a pity we do not know today which of the inns at Little Bytham was so named in 1854.

Some indication of the generous-minded spirit behind the five years of expense, trial and error at Grimsthorpe may be gathered from the following announcement in *The Times* of March 9, 1854:

'We are asked to state Lord Willoughby d'Eresby's steam plough is now complete and available for general use.

Any gentleman who wishes to see it in operation may do so on application by letter, two days previously, to Mr. Scott, Edenham, Bourne.

Lord Willoughby wishes it to be understood free permission is given either to take drawings of the machinery, or make any use that may be desired on the invention gratis.'

What more could anyone ask of a noble Lord, whose hand was wide open to the world!

In May 1855 Mr Scott wrote, 'We are ploughing by steam today and getting on very well', which I think may be accepted as the closing statement on the generous and great-hearted contribution made to steam work on the land by Lord Willoughby d'Eresby, who shortly afterwards turned his attention to a steam tramway.

A farmer named Hannam, living at Burcote near Abingdon in Berkshire, was one of the first to be associated with the introduction of what came to be known later as the 'Roundabout' system of ploughing or cultivating. This method took its name from the peculiarity of having a long free-running haulage rope carried over guiding pulleys, right round the outside of the field. The rope started from and returned to the winding winch, which stood near the single engine. Like some 5,000 other British farmers of 1850 Hannam had an ordinary farm type portable steam engine, i.e. one drawn from place to place by horses. As these engines were expensive items to purchase, and not always fully employed threshing, grinding corn or cutting chaff, Hannam had ideas about using their spare time to till the fields. With this end in view, he designed a windlass or winch (probably belt-driven), pulleys to guide the rope of twisted iron wire strands, and anchors to hold the rope straight while it was moving the implements. This equipment was ordered from Barrett and Exhall of Reading, and the ploughs most probably used were those ordinarily drawn by the horses, suitably modified for attachment to the cable. One great attraction of 'Roundabout' work was that it required only one engine, a feature much appreciated by the cash conscious mid-nineteenth century farmer, who turned each penny over twice before going in for what he called, 'This new fangled steam culture'.

Another farmer who made quite a name for himself by development of the roundabout layout, was William Smith of Little Woolstone, Buckinghamshire. He was keenly interested in applying steam power to land work, and started in 1853 by buying a portable engine and windlass before placing an order with Howards of Bedford for a light cultivator of his own design. By alternately reversing the direction of rotation of the two rope drums on the windlass, the cultivator was pulled to and fro across the field. 'Smith's System' became a farmhouse word during the next ten years or so; it was used freely to identify his particular adaptation of roundabout working as originally introduced several years earlier and improved upon by Hannam. Even though Smith's name is now but little remembered, credit must be given to him for his untiring efforts. Smith's tackles were usually made by J. & F. Howard, of Bedford, and there were no less than thirty-nine of them in use in 1858 at various places in England and Scotland. One can well imagine the consternation and excitement his steam machinery must have caused as it appeared for the first time in isolated country districts, with top-hatted squire, button-legginged farmer, and smocked labourer, together with boys, girls and women flocking around 'This last thing on Earth' puffing its head off in out-of-the-way fields.

Side by side with the latest advocates of rope, or indirect haulage, rose new promoters of direct haulage. In 1852, John Bethell invented a rotary digger in the form of a framework on wheels pulled by horses, but surmounted by a steam engine for driving the helical arrangement of digging prongs at the rear of the machine.

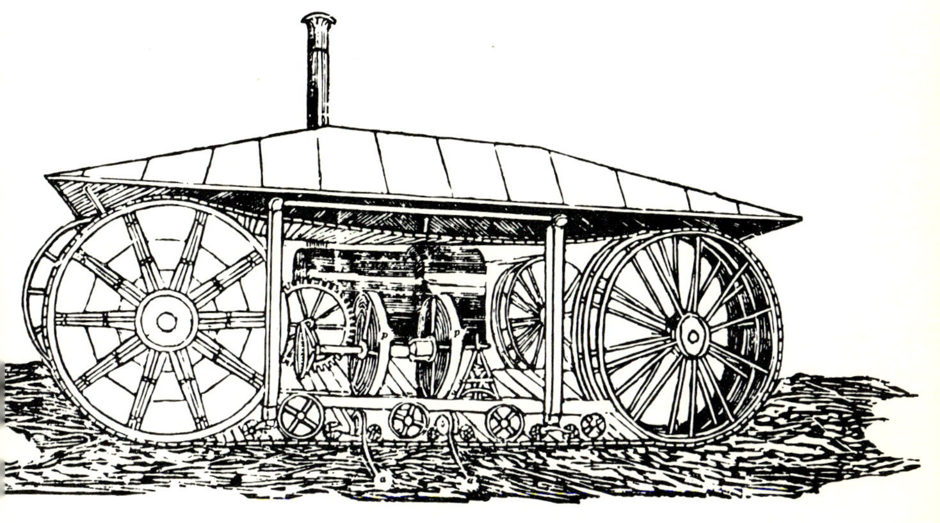

1. Heathcoat and Parkes' steam plough, 1833

[*Photo:* Patent Office Library

Smith's steam cultivator, manufactured and exhibited by Messrs. Howard, 1858. Roundabout style of work

[Reproduced by courtesy of R.A.S.E.

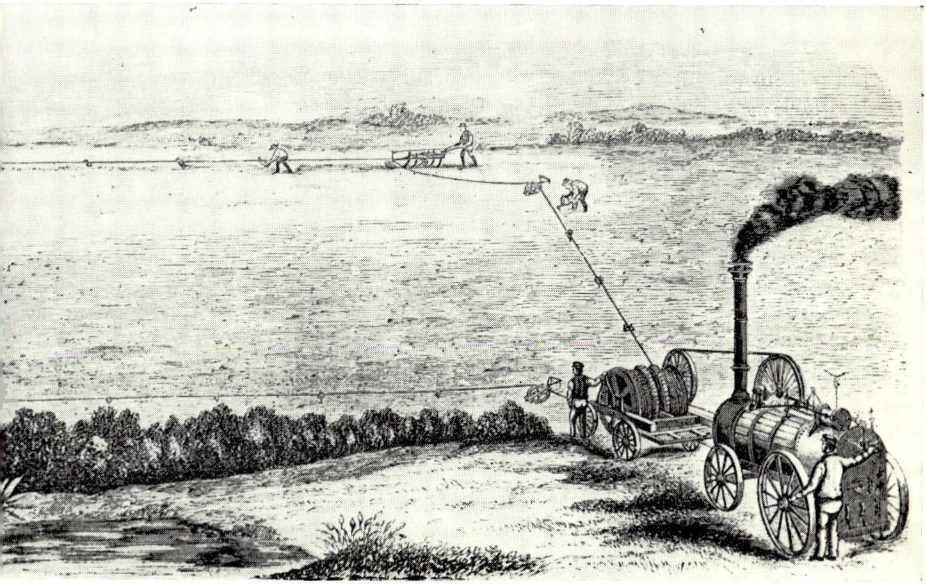

2. Fowler's patent double-cylinder plough engine with winding and clip type drums. Built by Kitsons about 1861

[Reproduced by courtesy of John Fowler & Co. (Leeds) Ltd.

Fowler's double engine set of ploughing tackle at work, 1880

Robert Romaine was a Canadian who came from Peterborough in Ontario where the vast and scarcely-tilled rolling Prairies still stretched westwards over thousands of miles. In imagination I like to think of him contemplating this scene with its challenge to agriculture so intensely, that he became convinced he had the answer in steam-powered ironwork as ultimately crystallized in the patent he took out jointly with John H. Johnson of London in 1853 after his arrival in England. Like Bethell's 1852 digger, horses were used to pull the implement up and down the fields, leaving the vertical boilered steam engine with its one upright cylinder to drive the spiked digging drum at the rear through a cardan shaft. Financial help was given willingly by the famous and progressive-minded agriculturist Jon Joseph Mechi of Tiptree Hall, Essex, who also encouraged Romaine through the difficulties and short-comings which cropped up during trials with the prototype digger. The extra expense and nuisance of restless horses, coupled with the great weight of the tackle which permitted work only while the land was fairly dry, as well as the high fuel consumption, all com-bined against practability. A second and improved horse-drawn digger financed by the Canadian Government was then built and exhibited at work near Paris during the 1855 Exhibition. This machine could obviously cultivate successfully because shortly afterwards the owners of the Crosskill Works in Yorkshire came forward readily to co-operate with further experiments. They built a new and an entirely steam-powered digger of formidable appearance, in their works at Beverley.

Leonce de Lavergne, a Frenchman who travelled here in the late eighteen-fifties, was an impartial witness of the widespread adaptation of the steam engine to British agriculture. The spread of steam cultivation impressed him more than anything else he saw in England.

IV

John Fowler

It is widely supposed that John Fowler was the inventor of the steam plough and while this supposition is not strictly correct, there is a good deal of truth in it. In addition to the many brilliant improvements he made on the basic ideas of others, Fowler was himself a clever inventor, and he has always been the most outstanding personality in the world of the steam plough. When he first became interested in power cultivation in 1850, he found only scattered pioneers working mostly alone, and often without any really effective workshop or equipment. Yet fourteen years later, at the time of his early death, he left the steam plough firmly established in production and practice. Fowler was a man who was usually right in his ideas, and he had also the enviable quality of co-opting others easily and happily in his experimental work. It was this ability to get on well with other people, combined with his personal zeal, which enabled him to build up such a splendid and enthusiastic team of experts. He got the right men together. I suppose it might be said of him that he was the great bright guiding star in a constellation of steam plough inventors.

John Fowler was born at Melksham in Wiltshire, July 11, 1826, of fairly well-to-do Quaker parents who gave him a good upbringing and education. His first job was in the corn trade, but he found himself unsuited to this, being of a mechanical turn of mind, and in 1847 he went to Messrs Gilkes, Wilson & Hopkins, a railway engineering firm at Middlesbrough. This company was then making locomotives, viaducts and bridges. While in their employ Fowler took a holiday in Ireland in 1849 and was horrified by the hunger and poverty that were the aftermath of the terrible potato famines 1846–47. When blight attacked the potato crops during those two years, it took away the staple diet of the country people, and so disastrous was the effect, that one and a half million inhabitants either starved, died of pestilence, just dis-

34

appeared or fled the country. Fowler at once thought he ought to do something to relieve the awful distress he saw everywhere among the labouring classes; more food was needed and the wet boggy land of Ireland required better drainage if crops were to yield better. As he thought about this, he recalled having seen somewhere a drawing of a mole draining plough, and thought it was just possible that this might be the very implement for the drainage job. Now a mole drainer is a sort of underground hole maker, or wheeled implement with a strong swordlike blade (or coulter) reaching down about three feet into the ground. At the tip of the vertical blade is a broader horizontal and bullet-shaped nose, which, as the implement is drawn across the field, leaves a mole-like, but slightly larger hole in the ground. Rain-water will then seep down the slits left by the knife-edged blade into the 'mole' holes below, and along them to the ends of the field where it can escape into the ditches. It is a cheaper, though less durable method of drainage than laying pipes in the ground. After thinking a little more about the mole draining idea, Fowler decided to leave railway engineering for good and dedicate himself to finding mechanical aids for draining and general cultivation. This decision was made for no other reason than to help provide employment and more food for the impoverished working classes.

His first drainer was exhibited at the 1850 Royal Show held at Exeter. The implement was hauled across the field by a fibre rope coiled on and off capstans whose bars were either pulled round by horses or pushed by gangs of strong labourers. Instead of just trying to make a 'mole' drain, Fowler hung behind the snout of the mole plug, a rope upon which was strung a long line of red kiln-baked drainpipes. As the drainer was moved over the land, it drew after it, down into the ground, the long line of pipes and so laid a very satisfactory pipe drain. The judges were much impressed with this smart piece of work, and no doubt it was one of the highlights of the Show.

As the Exeter method of draining had proved extremely heavy and exhausting work, Fowler tried his first steam engine in 1851. The engine he bought was really a big, heavy, portable type intended to pull the drainer behind it. Since this engine was not able to move itself it was drawn along by winding in a rope anchored at the end of the field, with of course, the drainer hitched behind. As may well be imagined, this proved to be a clumsy arrangement, and the engine made such heavy going of the work, that Fowler realized there was no steam engine then available suited to hauling implements. Reluctantly he realized that there was no alternative but to use a stationary engine.

His second steam engine was built for him by Clayton & Shuttleworth at their Lincoln works. It was a simple, straightforward type of portable, as used for threshing or other stationary duties, which was pulled from place to place by horses and used its own power solely for driving a belt connected from the flywheel to the various pieces of machinery. Fowler made up a set of mole draining tackle for it, comprising a two-drum windlass, anchors and pulley wheels, and a long iron wire rope which was led out along two sides of the field, roundabout style. This drainer machinery was exhibited at the 1854 Lincoln Show, where it worked very satisfactorily. Among the many spectators who were deeply impressed by this successful performance of steam power, was William Smith of Woolstone, already mentioned. He went away determined to renew his own experimental work.

It is not surprising to find that after the good show put up at Lincoln, other capable people, who believed there was a future for steam in cultivation work, soon associated themselves with Fowler. They recognized their leader. There had, of course, been meetings and discussions earlier because there is the record of a two hours' stroll along the pebble beach at Brighton one lovely summer's evening in July 1852, when Fowler and William Worby had pondered over the possibilities of ploughing by steam. Apparently the pair of friends talked 'shop' instead of enjoying their seaside holiday in complete relaxation. However, in spite of the bracing air of Brighton, and with full knowledge of the earlier 'near misses' by other pioneers, they came to the conclusion it would never be practicable to use the steam engine for ploughing.

There was at this time a farmer named David Greig living in Essex. He was a red-bearded Scot bursting with initiative and new ideas for power farming. He also became friendly with Fowler and Worby, and it was he who put forward a strong claim that the balanced type plough could provide a solution for the 'no turning' requirements of cable work. It was not his invention, because a similar plough had previously been used with horses. Briefly, this type of plough consisted of two opposite and inwards-facing rows of ploughshares, balanced see-saw fashion over the centre axle.

One set of the shares was right and the other left-handed, so that although the plough moved backwards and forwards across the fields without turning, all the furrows were laid in the same direction with crests of even spacing. All that was required at the ends, was to lift the side of the plough which had just completed its line of furrows, out of the ground; and pull the opposite end shares down into the soil, for the return journey.

During 1855 a change of outlook seems to have taken place, because in January 1856 the three friends had narrowed their debate down to the choice of ploughs. Worby advocated the long established Kentish 'Turn-Wrest' type on which the alternate direction ploughshares were tippled over on a horizontal axis in line with the direction of ploughing, but the choice finally fell upon the balanced plough recommended by Greig, which he had first seen used with Fisken-designed tackle.

In January 1856 Fowler addressed the Royal Society of Arts in John Adam Street, London. His talk provoked a good deal of questioning about the real prospects of ever being able to plough by steam. In the audience was Thomas Atkins, an Oxford civil engineer, who stood up and told the meeting that already he had spent £2,000 and more on such experiments since 1843. In Atkin's opinion an engine of not less than 40 h.p. was necessary to handle satisfactorily ploughs or other agricultural implements. Worby, who sat next to Fowler, said he was sure a 6 h.p. engine could be made to plough an acre per hour on light land. When he heard that, Fowler quickly replied, 'If you believe this, I will give you an order home with you to make a set of tackle at once'. The two friends stayed that night at the Great Northern Hotel, King's Cross, where the final bargain was made. Early the following morning they both left for Hainault Forest, where Fowler had a clearance and drainage contract in hand. After looking at the steam engines there, Worby set off as fast as he could for Ransomes & Sims works at Ipswich; he was determined to lose no time in building the promised steam plough. I think, too, that he felt he was on pretty safe ground, perhaps because he had already tried out a cable-hauled plough without using a steam engine.

There must have been some long hours and late nights in the workshop, because by April 10th that year the tackle was set up and at work at Nacton in Suffolk. It consisted of a Ransome's portable engine, a double drum windlass, long iron wire rope, rope anchors and a pole-steered balance plough. To everyone's delight, the thing worked. The promised acre an hour was accomplished quite easily. What a proud moment it must have been for the three chief promoters! The only misgiving they seem to have had was the amount of labourers' work necessary to dig holes for the shifting positions of the rope anchors, as the ploughing advanced down the field. Shortly afterwards the Nacton type of ground anchor was replaced by a four-wheeled anchor cart. This cart was loaded with earth to give it weight, and the wheels were of the sharp edged disc pattern, which cut down into the soil, and so resisted the inwards pull of the rope. It was far easier to pull this cart along than to use a gang of labourers continually digging holes for anchors.

In 1856 Fowler patented a system for ploughing with two engines, using one engine on either side of the field. I like to think that he had seen and accepted Lord Willoughby's invitation in *The Times* of 1854 to inspect the ploughing engines at Grimsthorpe. If he actually did so, he must have noticed the advantages of a twin-engine system. With two plough engines working as a pair, there was no need to use horses to move the rope winches, or to pull out the long rope before making a start. These self-moving twin engines of 1856 were taken to the 1857 Show at Salisbury, but at the last minute Fowler decided not to give an exhibition of ploughing with them. Two possible explanations are that it took eight horses to help pull one engine up a steep zig-zag hill to the trial ground, and owing to the domed shape of the field, neither driver would have been able to see the other engine. Instead, he showed an 'improved' single engine outfit. The engine and anchor cart of this tackle were each able to move themselves forward along the field headlands by winding in a rope anchored ahead. But once again the £500 prize offered by the Royal Agricultural Society of England for a satisfactory steam plough was withheld; the judges considered the steam work at Salisbury was no cheaper than if done by horses. In spite of this disappointment, however, there must have been some satisfaction at the excitement the steam machinery caused among the spectators. One may well imagine that Fowler and his engineers were bound to be the centre of attraction. The curious and cynical alike must have pressed around them with a bombardment of questions. Those who placed orders for new tackle were, of course, the most welcome.

In the late 'fifties there was an upsurge of orders and the sales of ploughing engines and implements increased to such an extent that Fowler was obliged to look round for fresh firms to build for him. One fact which must be remembered is that so far Fowler was not a manufacturing engineer himself; all his machinery was made by other firms. Engineering concerns like Ransomes & Sims of Ipswich, Stephensons of Newcastle, Kitson & Hewitson of Leeds and Clayton & Shuttleworth of Lincoln were all doing work for him at that time. The price of a one-engine roundabout set of tackle in 1858 was £508, but even so there were plenty of buyers. It seemed as if a great future was assured for steam machinery as soon as the rooted objection to a change from animal haulage had been removed from the conservatively minded English farmers.

Of the many snags of the early 1850's, none was more annoying than the clumsy chains used by Lord Willoughby, or the soft and easily kinked iron wire ropes in more general use. Steel was still a scarce and expensive metal, but Fowler knew that the only satisfactory rope would be made of steel. In 1857 he went to see a Birmingham rope maker and put his views with such force that the same year a steel rope was actually produced. It was a revolutionary improvement; the life of this rope extended over a thousand acres of ploughing compared to about 200 acres with an iron wire rope. From that day to this, high quality steel wire rope has been classed as 'Best Plough Steel', as any maker's initial test certificate still shows.

The next logical step was to devise a satisfactory, self-moving plough engine. The farmers were getting a bit tired of taking out a team of horses whenever a move about the field, or from field to field, was necessary. The backroom boys at Fowlers had been busy on this possibility for some time, but it was not until 1858 that they turned out their first practical self-moving traction type plough engine. As was usual in those days, it required a horse in shafts at the front to steer it. Where the horse was led, or went, the engine followed. I suppose we find it hard today to understand why so simple a matter as steering gear was not perfected earlier. Perhaps the engineers were so fully occupied with other design work that steering mechanisms had to wait. One cannot help feeling sorry for the poor horse; it must have been exceptionally docile or well trained, to put up with the frightening ordeal of having a puffing and clanking steam engine right on its tail. Incidentally, sometimes these horse-steered engines made fairly long journeys by road. There is, for instance, the record that Mr J. A. Williams returned home from the Salisbury 1857 Show some 40 miles to Baydon, Wilts., driving his Clayton & Shuttleworth portable engine converted to chain drive, guided by one horse in shafts.

Another highlight in John Fowler's career was the £500 prize he won at the 1858 Chester Show. This show was held on the wide and flat 84 acre Roo Dee field which lies in the arms of the Dee, and over which Roman galleys were rowed when the Dee flowed closer to the old town walls than it does today. On this occasion Fowler showed the beautifully proportioned steam plough engine built for him by Stephensons. It was a golden day for him because he won the prize for proving steam ploughing was a cheaper alternative to horses. The discerning judges, who had withheld the £500 award for so long, now thought fit to admit the case was at long last proved by the competition ploughing John Fowler carried out on a farm just outside Chester.

In spite of winning the Chester prize, Fowler still needed lots of money to continue his expensive steam plough experimental work. He resorted to the unusual method of forming, in 1859, 'The Steam Plough Royalty Co. Ltd.', in which his relatives and friends invested in the hope of picking up royalties in the years to come from Fowler's numerous patents. This venture is said to have been one of the earliest known joint stock companies. While mentioning financial backing, it must be borne in mind, too, that no little assistance was given to Fowler by large land-owning gentlemen; friends of his who were quite naturally interested in maintaining the value of agricultural land. They thought steam ploughing would make farming more profitable and so push up the price of their land.

In 1861, Fowler fitted Clip or Burton type rope drums, as used in many collieries for cage winding on some of his Kitson built plough engines. The hinged jaws on these drums closed on the rope under stress and held it firmly in their grip, without any need to pass the rope more than once round the drum. Of course, the success of these drums depended upon the slack gear on the ploughs keeping the rope taut. We must bear in mind that this type of drum preceded the coiling one.

Fowler's business grew by such leaps and bounds that he outstripped the capacity of the firms who built for him under contract. By 1860 there was no alternative but to start his own works, and that year he opened his 'Steam Plough Works' at Hunslet, Leeds. His first really satisfactory two engine set left these works in 1861. Another important event was the patenting, in 1863, of an automatic gear for coiling the ropes on the engine drums. Briefly, this apparatus, which was incorporated inside and below the drum, consisted of two toothed wheels, one above the other revolving round the drum centre shaft. The number of teeth on these wheels depended upon the number of rope coils on the drum, and they were meshed with a small free running compound toothed pinion. But since the top or clutch wheel had twenty-five teeth for twelve coils of rope, while the lower or tappet wheel had twenty-four teeth, a slow relative motion was imparted to the lower wheel. The motion of the bottom wheel moved the arm receiving the rope slowly up and down at the rate required to coil the rope correctly and neatly on the drum. Before these coilers, known as reverted epicyclic gear trains, were invented by Fowler there had been a great deal of trouble with roughish winding. The ropes would lap irregularly over preceding coils with frightful jerking and bumping, and of course, ropes so ill treated soon broke.

By the end of 1863, it could be said that at last steam ploughing was established on a paying basis. Unfortunately, Fowler had overworked himself and was now in a poor state of health. He took his doctor's advice to move and lived in the little village of Ackworth, 12 miles outside Leeds. Horse riding was suggested to him as a suitable form of exercise and he often rode in to his office. It was while out fox hunting in November 1864 that he fell from his horse and suffered a severe compound fracture of the arm. In spite of the best available medical attention, blood poisoning set in, and in his already weakened state of health, he could not withstand it. Three weeks later, at 5 p.m. on December 4, 1864, he died, aged only 38. He left a wife, formerly Elizabeth Lucy, ninth child of Joseph Pease, M.P. (one of the promoters of the Stockton & Darlington Railway in 1825), and five children. At the time of his death, no less than 700 men were employed in his Leeds works and were turning out one engine weekly.

The Pease family erected a red granite memorial to Fowler in the South Park at Darlington. It is surmounted by a bronze model of a three-furrow balance plough and the simple inscription reads,

'John Fowler 1856', the year in which his first successful steam ploughing took place at Nacton. So far as I know, this is the only monument erected to his memory. His name has, however, been taken round the world on the maker's brass plates attached to the many thousands of Fowler traction and plough engines, subsequently built in the Steam Plough Works.

At the 1963 Annual General Meetings of both the Road Locomotive Society and the National Traction Engine Club, I proposed we should commemorate the 1964 Centenary Year of John Fowler's death by erecting a national memorial. It is now likely that a sculptured plaque will be presented to the new College of Agricultural Engineering at Silsoe in Bedfordshire.

V

Oxman and Horseman to Steam Ploughman

In order to understand what was involved in recruiting steam plough crews from agricultural workers, who till then had been engaged almost entirely in the care of animals, it will be helpful to take a look at the general conditions of farm workers in the mid-eighteenth century. As we have seen, wages were wretchedly low, yet even so, the men managed to drink large quantities of ale which they often bought direct from the brewer, in 18-gallon casks, at 6d., or beer at 1/– a gallon. In harvest time, work went on until ten or eleven at night without any extra pay, although the farmer usually provided free soup and beer during the extended hours.

As the reaper had not yet replaced the scythe and sickle, harvesting was a lengthy and extremely laborious affair. In consequence it was not until all the corn had been safely carted and stacked in the farmyards, that either men or animals could be spared for the autumn ploughing. October was usually well advanced before a serious start was made to break up the stubble.

Oxen were still often used in plough teams, plodding along sleepily at not more than $2\frac{1}{2}$ miles an hour. Mr. E. Ruck, Castle Hill, Cricklade, had no less than fifty-six draught oxen on his farm. These animals were made up into teams of four; seven for morning and seven for afternoon work. Both ox and horse ploughmen wore the common dress of a coarse cotton smock.

It was the custom to afford a measure of relief and rest to man and beast by giving a long break from field work at Christmas. The teams laid up until the first Monday after Twelfth Day, known as Plough Monday. At the time steam ploughs were introduced, and long afterwards it was a widely observed custom on Plough Monday, for the ploughmen to take a plough from door to door, collecting 'Plough Money'. The cash they got by this means was spent in a 'Booze Up' before recommencing ploughing in the frosty or muddy fields after their Christmas holiday.

Sometimes, in an effort to make the going easier, the plough-shares were set craftily on the shallow side, or the ploughmen shirked holding down the swing type plough stilts (handles), with the result much tillage was shallow and crops light in yield.

From a survey made in 1857 it was reckoned that four oxen or three horses ploughed, on average, four acres a week. One interesting advantage the steam people claimed over horses was that the cable-hauled steam implements did not pack down the soil so badly as the hooves of horses drawing ploughs. It was assessed that there were 300,000 horse hoof marks per acre, so perhaps there was some point in avoiding all these impressions. Another important economic fact was that on completely animal-powered farms, the draught beasts ate grass, hay, straw or corn to the tune of one-quarter of the whole farm's produce. Further, since these animals fed for long period before and after work, it was necessary for the ploughmen to toil early and late feeding them in their stalls. Early records show, how in 1867, young horsemen worked from 4.30 a.m. till 7 p.m. during the summer months, in the East Riding of Yorkshire. The day's work was similar in winter in Lincolnshire. What the chaps there called 'Suppering Up' occupied them from six to seven of a winter evening, when the wooden racks over the mangers were piled high with clover hay, wheat or barley straw for the night-long munching of the heavy plough horses. At 4.30 in the morning the shire horses were ready to begin eating again as soon as the morning's fodder was brought to them. Over a large part of England at that time unmarried horsemen were engaged on a yearly basis and paid about £11 to £12 for the twelve months. They were boarded out by the farmer on the waggoner's wife, who fed them cheaply on skimmed milk, cow beef, coarse and often over-salted home-cured bacon, and home-baked bread.

From all this it is obvious there were plenty of farm workers well used to long hours of hard work. Many of these men were therefore willing to try their hand as steam ploughmen for the sake of earning a few extra pence a day from acreage payments. Others objected strongly to the coming of the steam engine; they feared it would not only oust their cherished skills with rein and plough stilts as they drew dead straight furrows, but also put ploughmen out of work at a time when a job of any sort decided whether a man and his family starved or not.

The first steam engines seen by many country labouring men were the portable or horse-drawn types introduced from as early as 1832 by the more mechanically minded and progressive landowners. These engines were not employed for ploughing, but for threshing or grinding corn, pumping water, cutting chaff and other stationary jobs. It has been estimated that by 1851 there were 8,000 of these engines at work on British farms. The extension of the railways in the period from 1825 to 1860 gave others working by the linesides a daily acquaintance with the puffing and clanking railway locomotive. The result was that here and there, this or that boy or man was attracted to steam; it got into his blood. To a large extent the first skilled enginemen on the railways were obtained from the North of England where the earliest large scale locomotive manufacturing works were established. In a smaller way this recruitment of experienced men from the north was applicable to steam plough engines. Fowlers sent a trained man with all new tackle they sold, to instruct the local men in the handling of engines and implements. Very often the purchasers offered the firm's men sufficient inducement to stay on. It then fell to them to teach men who could neither read nor write, the rudiments of driving steam engines, or steering the plough and cultivator.

The recruitment of satisfactory crews was a major consideration of the new steam plough owners. There was a great deal of discussion about the problem at meetings organized by the Royal Agricultural Society of England (subsequently referred to as the R.A.S.E.). Opinions differed, as may be expected, but on the complimentary side Mr E. Holland, M.P. of Dumbleton, Evesham, told the Society in 1862, 'It was an unexpected and gratifying fact that our labourers have turned out to be most excellent and efficient engine drivers when once initiated by a mechanic who knows his duty'. Advice on choosing men was given by a Mr Amos who said, 'Select a steady-going labouring man, clean and particular in his habits, who would feel himself flattered by the trust reposed in him'. Later on it was remarked at a lecture given in 1867 in the presence of several noble lords, 'We have a new generation of young steam labourers, a species of village aristocracy who value their position'. Sometimes it so happened the village or estate blacksmith became the engine driver; and one may expect this to have been as good a choice as any. Engine driving could often be a bit of a nightmare. Boiler feed pumps, safety valves or steam gauges were notoriously unreliable, and in consequence boiler explosions fairly frequent. Sometimes when the plough chanced to strike a 'landfast' stone or tree root, the driver would open out with all the steam he had to release the implement. Quite frequently this 'burst of power' either broke the rope, or the anchor was pulled out of the ground and sent flying. Some of the old farmers of the anti-steam brigade were all too fond of the jibe, 'Now then, gently over the stones!'

Anyway, in spite of teething troubles, the fact remains that on the whole the idea of ploughing and cultivating by steam began to be accepted as a normal, everyday affair on the farms of this country, and there were plenty of men prepared to man the gangs.

VI

The Royal Shows

The Royal Agricultural Society of England was formed in 1839. In its journals of the next fifty years or so are to be found a wealth of articles and illustrations concerning steam cultivation which, at that time, was a fascinating topic in agricultural circles.

The Society also played a most important role by providing at its yearly shows a convenient stamping ground for exhibitors of all forms of farm steam engines, especially those intended for cultivation work. As these shows moved from place to place about the country, they presented an unequalled opportunity for the largest possible number of interested people to see what was going on. The general scenes must have been outstandingly colourful. There was for instance the avenue of 112 long chimneyed steam engines of various kinds at the 1858 Roo Dee show, each with its gaily painted flywheel turning in the sunshine. Visitors' dress was equally gay with the ladies in crinolines, while the gentlemen wore red or yellow waistcoats and, of course, top hats. Labouring men usually wore long white smocks patterned front and back with decorative smocking.

In some instances the engines exhibited at the shows had been hastily constructed in newly established or poorly equipped workshops. This was due to the universal rush to cash in on individual ideas; nobody wanted to miss the boat. This led to some disregard for safety, particularly with boilers, which although only pressed to about 45 lb, were often badly constructed and dangerously inclined to burst. At the 1858 show no less than three engines had to be thrown out due to risk of boiler failure. As a safety measure the R.A.S.E. had to insist that their judges should inspect and certify all boilers and engines were in a satisfactory condition before they were allowed to take part in any tests. Unfortunately, of all that great and varied assembly of early cultivating engines, whether they were exhibited at the shows or not, nothing has

47

survived. It is only in imagination, working on descriptions or early engravings, that we can attempt to recapture what they looked like in the days when the steam plough was but an infant wrapped in the notions and ideas of hundreds of patentees.

We must remember, too, that nowhere else in the whole world could those stirring steam sights be seen. The engineers of England had conceived the steam plough. If foreigners wanted to see this wonderful new machine of great promise to agriculture, they had to sail to this island. Many came and marvelled; moneyed land-owners or rulers of countries were so impressed that they gave orders for tackle to be exported for use on their home estates. Those were England's superb days—she led the world in steam machinery—and particularly power cultivation. The introduction and establishment of steam power on the land and elsewhere was a tremendous stimulus to subsequent higher standards of civilization. Our contribution in this respect has yet not been sufficiently acknowledged by historians. It is something about which we can be proud.

Another important feature to remember about these shows is that the steam plough exhibitors put over high pressure sales-manship. The established firms like Fowlers, or Howards of Bedford had their engines specially built and tuned up for the occasion. Their steam plough gangs were formed from selected and highly trained men and boys who worked like whipped horses under the spur of bonus payments, or extra beer. Deficiencies in the design or performance of the tackle were often covered up by smart team work on the part of these shock troops. No ordinary purchaser of steam cultivating machinery could ever hope to have the benefit of such capable crews. There is the recorded occurrence of how at the 1858 Chester Show one over-zealous engine driver, working his engine under test conditions, was noticed to be appropriating a handful of cinders from along-side another engine in order to keep his own dying fire alight. He was promptly listed as a 'culprit'.

Rivalry was very keen among the competitors. At the Newcastle Show of 1864 a John Fowler and a Howard set of ploughing tackle ran what is described as a 'neck and neck' race in adjoining fields. Even the judges disliked the fast pace of the Fowler set, and not without foundation because a little later one engine flywheel flew off and landed in a hedge, fortunately without causing any

3. 1856 Burrell Engine with Boydell 'Endless Railway' wheels

Savory's Drum-round-the-Boiler Ploughing Engine of 1861

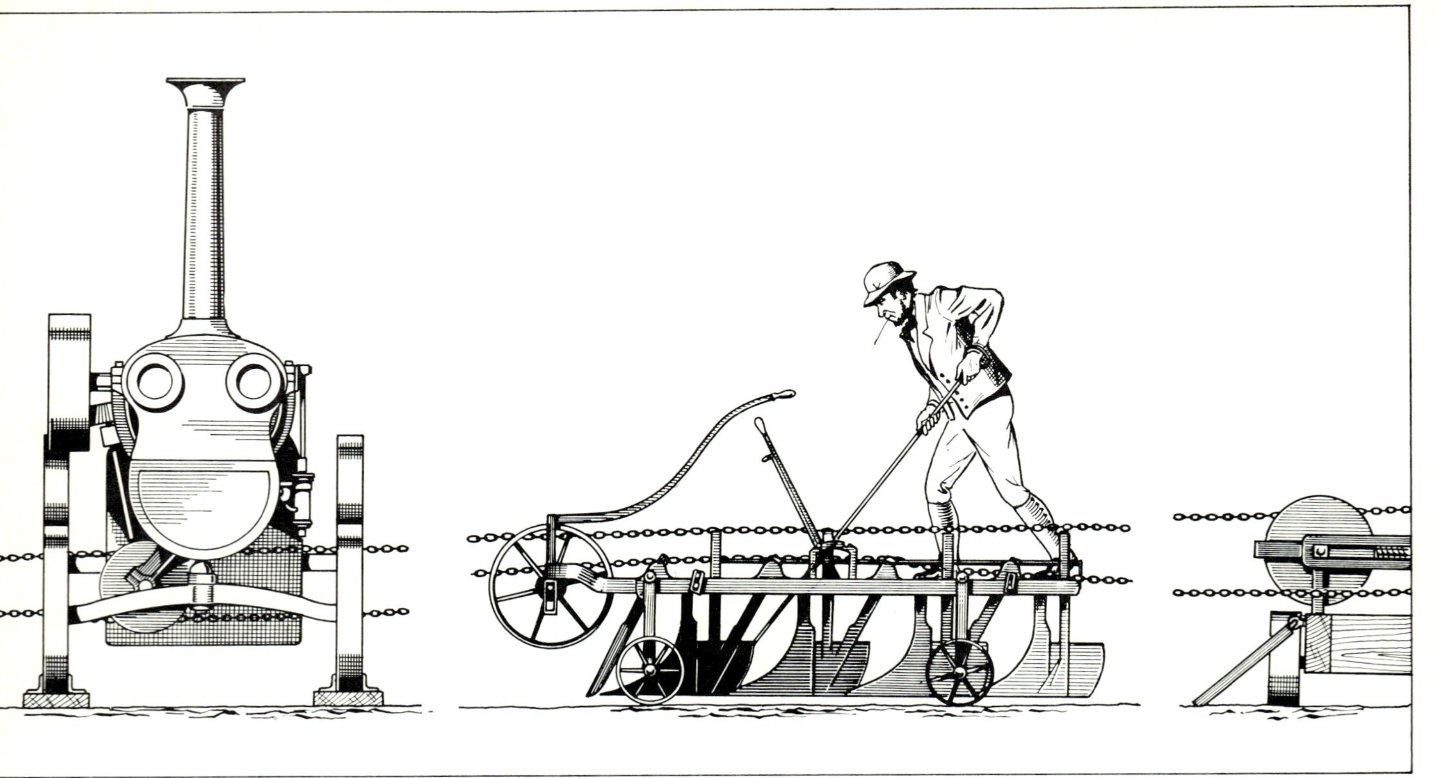

4. Lord Willoughby's Steam Plough of 1850

personal injuries. Howards likewise flogged their engine so hard that the smokebox and chimney became red hot. Their implement stuck on a big stone and threw and injured the man riding on it. Eventually, with mounting excitement among the spectators who began to crowd in on the wire rope, the trials had to be stopped. A point of some interest is that on this occasion, the Fowler tackle comprised a set of two engines. Each had a clip or grip type drum and the endless rope ran round the drums of both engines just like a belt over two pulley wheels. The object was to make each engine pull on the rope all the time, so that in effect the implement was hauled by the joint power of the two engines. If this method could have been made practicable, it would have avoided the criticism of ordinary two-engine systems, that only one engine at a time was working. However, in practice it required close synchronization starting and stopping the pair of engines. The judges awarded Fowler the prize, but they had no high opinion of his 'Twin System' which they thought impracticable. The uneconomic feature of one engine idle half the time and its driver also idle at the same time was never overcome, and weighed heavily against steam ploughs in later years when tractors handled by one man came into open competition with the two-engine ploughing sets.

After 1854, when Fowler brought the first steam mole drainer to Lincoln, steam cultivation machinery exhibits were the principal attractions at the R.A.S.E. Shows. By 1871, however, it could be said the novelty and excitement of the steam cultivating exhibits had worn off. All the same, the Society had done a splendid job in allowing the steam plough engine to show its paces over such a wide field.

VII

Other Notable Pioneers

'Many are called, but few are chosen', so runs the apt phrase in the Bible. No exception was granted to the steam plough pioneers, the vast majority of whom toiled hard and often well without achieving a place of fame in posterity. One of the first things which impressed me when I began my researches for this book, was the multitude of men in all walks of life who 'had a go' at inventing something or other in the way of steam cultivating machinery. The remark made by Robert Stephenson (son of the famous George Stephenson) in 1841 that 'The steam locomotive is not the invention of any one man, but of a nation of mechanical engineers', applies with equal force to the steam plough. Unfortunately, it would be quite beyond the compass of this book to do justice to all individuals so only a limited account of the more famous will be given.

W. Fisken was a Presbyterian minister at Stamfordham, Northumberland, who found time between preaching on Sundays and visiting his flock on weekdays, to apply his inventive flair to finding means of using steam engines to till the soil. He set to work in 1852 with the aid of a money grant from the Highland and Agricultural Society of Scotland and was able to produce for exhibition at the 1855 Carlisle Show a single engine, roundabout style, ploughing tackle. Now the bee in Fisken's bonnet was a belief that a long, endless and fast running hemp rope driven from a groove in the flywheel of a stationary engine was the best way of transferring power to the implements. Although it was certainly a possible method, and was subsequently improved upon, the speed of the scurring rope ($12\frac{1}{2}$ times that of the implement) raced the supporting pulley wheels so much, that their bearings ran red hot in no time. Further, the intermediate gearing on the anchor cart installations, or in some instances on the drive mechanisms on the four-wheeled equipoise or balanced plough, absorbed too much of the available power, and during wet weather the hempen rope must have shrunk, and got muddy and bedraggled. Even though the enterprise was later taken over by the Ravensthorpe Engineering Co. which manufactured a fair number of sets after 1870, Fisken's venture has to be classed with those which 'also ran'. Perhaps Fisken's greatest contribution to steam ploughing was his balanced plough which had so impressed David Greig, and which he persuaded Fowler to adopt in 1856.

A unique method known as 'Halkett's Guideway System' was designed and introduced in 1855 by Lieut. P. A. Halkett, R.N., a landowner in Wandsworth. He was another inventor with rather grand ideas, which were limited in scope. Although his system with two engines running on parallel rails or guideways, connected with a 50-ft girder bridge from which the varied implements were suspended, was spectacular enough; it was restricted to the small areas for which the extensive ground equipment could be afforded. Such functions as ploughing, drilling, reaping, manuring, underground watering (irrigation delivered just below the surface), harrowing, subsoiling and even working at night by lamplight were contemplated. There was also the associated idea of a guideway leading to the farm buildings in order that one engine might go home as and when required, for barn work or threshing. Extensive trials were carried out but did not give sufficient encouragement for Halkett to proceed beyond the prototype.

When the earliest trials were made with steam engines pulling their own implements, it was at once obvious that one of the main faults was the way in which the wheels sank down into the ground. It was thought by some that if the weight could be spread over a larger surface by a different type of wheel this difficulty could be overcome. James Boydell thought a wheel fitted with five hinged flaps was the answer, and while his reasoning was right, it was too early yet for suitable mechanisms to be manufactured. The flapping, ducklike feet clanked terribly, and the hinge bearings rapidly wore out.

In 1861, two young friends, Alfred Fernandez Yarrow and James B. Hilditch, spent much of their weekends or holidays in a little workshop at Woodlands, Blackheath, experimenting with various electrical, photographic, steam or other scentific projects. At 17 these bright young boys made and installed a telegraph between their two houses half a mile apart; this was probably the first telephone system installed in London. At one period, cultivation by steam caught their imagination, and they planned a one-engine type of tackle aimed at eliminating the usual heavy pull on ground anchors, which were often uprooted and overturned on other systems. In their layout, two cultivators worked to and from the centre of the field, each keeping to its own half and always working under load as it moved inwards towards the one engine. This meant there was always a direct pull from the engine drum when the implement was in its 'working' position, and the comparatively light haulage of the 'empty' outwards runs was the only strain placed upon the manually-moved opposite anchor. The need to stop and reverse the engine when the implements reached the middle of the field was a drawback, as this slowed down the rate of work.

The implements for their first set of machinery were manufactured by Coleman & Morton of Chelmsford. This tackle, powered with a two cylinder, side drum, modified Clayton & Shuttleworth traction type engine, was shown at the 1862 Royal Show. It was bought by R. & T. Wagstaff, Stifford Clays, Grays, Essex (potato merchants in Spitalfields and King's Cross markets), who said that the engine was light on repairs, and satisfactory in every way. Later, no less than twenty-four Yarrow & Hilditch style cultivating outfits were sold to the Khedive of Egypt for the cultivation of cotton land; a direct result of the American Civil War of 1861–65 had been a greatly increased demand for Egyptian cotton. Incidentally, Yarrow afterwards turned to marine engineering and is still famous for the water tube type boilers which bear his name.

W. Savory & Son, High Orchard Iron Works, Gloucester, brought out about 1861 an unusually interesting ploughing engine which may have been built for them by Robey & Co. of Lincoln. It had the distinctive arrangement of a shell style rope drum completely encircling the boiler. Power was obtained from a two cylinder short stroke, high speed 10 h.p. engine fixed transversely across the smokebox, in a position where it could conveniently drive the crankshaft which passed back inside the drum. The rope drum and the driving wheels were both driven directly from this longitudinal crankshaft.

Some engines constructed from 1863 onwards at the Gloucester Works had inverted vertical type cylinders and did good work as twin sets in the Worcester and Gloucestershire areas, but the Savory idea never caught on generally. One set is reported as having travelled 30 miles in a day; not bad going when one considers what the state of the roads must have been at that period. There is even an unconfirmed legend that one of Bomford's Savory engines travelled under its own steam to the 1864 Show at Newcastle. In 1864/5 Garretts of Leiston may also have built two pairs of these drum round-the-boiler Savory engines. Although the idea of using a high speed engine was sound, to my mind these small size ploughers look as if they would not have been strong enough for anyone but a light land market gardener.

Max Eyth was a young German engineer who had heard of the great activity in steam engine construction then taking place in England. In 1861 he decided that if he wished to make his career in this line, the best experience he could obtain was in England. He crossed the North Sea in the same year. His role is a little different from the others in this chapter because his fame was not as an inventor on his own, but as one of the most outstanding engineers in John Fowler's team of experts. As soon as Eyth arrived in this country he wrote to Fowler enquiring about employment. After a time he received the following interesting reply:

Dear Eyth,

My friend Tyler in London reminds me about you. If you are inclined to commence work in my factory, just recently started, you will find a vice. As soon as opportunity offers, I will take care that you learn steam ploughing. After that we must see. I believe in the future of the thing. To begin with I offer you thirty shillings per week: with this you can live and for the present it should suffice.

With kind regards,
Yours truly,
John Fowler, jun.

Eyth gladly accepted the opportunity to enter the new Steam

Plough Works at Leeds, where he soon showed promise. Besides being a skilful and imaginative draughtsman and a prophet on the prospects for the steam plough, he was also a wonderful letter writer, so much so in fact, that letters home to his parents in Germany were long a classic for engineering students in German high schools. I like the way he describes John Fowler for whom his respect and affection almost amounted to hero worship: 'A splendid man about 34, big and stately, black haired and affable, with a laugh that did good to all within a hundred yards of him.'

It was not long before Eyth was out testing tackle, for he tells how while working at Hadham, Hertfordshire in 1861, Fowler travelled down from his office in Cornhill, London to see the machinery in action. When he entered the field he was at once very pleased with the way things were going. He noted Eyth had extremely dirty hands and there and then, in addition to a few kind words, gave him a £10 advance on his wages.

In 1863 Eyth was a Fowler representative in Egypt where at that time the rulers were gripped with steam plough fever. Apparently they used to have some stirring scenes out there, too, as for instance at the demonstration ploughing at Shoubra, near Cairo where work continued into the night, aided by torches and moonlight. The whole affair amazed the Egyptians and even Eyth thought it weird, as though demons were at work. He was in Egypt when news of Fowler's untimely death reached him. Commenting on this he wrote, 'There dies one of the noblest, most talented and capable men with whom I have ever become acquainted'.

The Third Duke of Sutherland, fired by wonderful visions of what might now be done with the aid of steam, undertook in 1872–78 a big and ambitious scheme to make some of his poorish moorland at Strath Terry, on the NE shores of Loch Shin, grow oats and root crops for winter sheep feed. Eyth was there with the eight Fowler steam plough sets and seven traction engines working on the project. The Duke, who had some engineering ability, had designed the special plough-cum-stone-upender known as the Sutherland Reclamation Plough, with its rather fearsome single tines dubbed, 'The Duke's Toothpicks'. Once again we have a lovely bit of prose from Eyth, who apparently delighted to take a pen and paint a word picture of the sights he saw:

'An immense hook, like a big one-armed ship's anchor, was drawn two feet deep through the soil by 30 h.p. engines, having an actual horsepower of 80, and tore out of the ground all the stones which lay in its path. Blocks of half a cubic yard seemed to appear playfully from the underworld. If the implement came to an old granite drift block, not to be overcome in this way, dynamite cartridges were exploded under it, which the workmen always carried with them, carelessly, along with their bread and bacon. After this preliminary operation, the perfectly white field, covered with loose rocks and stones, had the appearance of a stormy sea solidified. It would have been impossible to drive the roughest wagon over this stony sea without demolishing it. A skilfully constructed sledge was then drawn backwards and forwards between two ploughing engines and on it the stones were pulled to the sides of the field. There the sledge automatically overturned and threw its fearsome load off. Thus along the field borders were formed high ramparts of boulders which were used partly for building houses and stables, but also for substantial walls enclosing the fields themselves. Only then could an ordinary steam plough begin its work and the field be ploughed correctly for the first sowing with oats, which with house and farm buildings would be given to the new tenant.

Two thousand acres were thus improved at Strath Terry as well as 500 more at Kildonan.

In Germany they still credit Eyth as the inventor, while working at Fowlers, of the second and more successful type of automatic rope coiling gear so universally used to the end. This may be partly true because even though the first rope coiler was patented by Fowler in 1862, the second and more effective type of about 1875, incorporating an eccentric gear wheel inside the drum, does not appear in the Patent's Office lists. It may well be, therefore, that Eyth invented this mechanism, but as a German national did not attempt to apply for a British patent. At last a longing for home seems to have come over him, and in 1882 he returned to Germany with the firm purpose of applying his English steam ploughing training and experiences to the promotion of a similar industry in his homeland. It was perhaps due largely to his efforts that Germany became the only country where two-engine steam plough sets on the style set by Britain were built in any quantity. The engines made by some German firms had a decidedly Fowler-like appearance. Eyth also imported into Germany the idea of a national agricultural organization, similar to the Royal Agricultural Society of England. In German agricultural history his name is honoured as one of the 'Greats' of the Fatherland.

This diversion on Eyth has, however, taken the story a little ahead of time, so something must now be said about the success or otherwise of steam ploughing in the two important decades following Fowler's death in 1864.

VIII

Iron Horse v. Hided Horse

By the end of 1864 it could be said steam cultivation had just about established itself as an economic aid to agriculture in Great Britain. The opinion expressed by a Hampshire farmer in 1861 about steam plough contractors, 'Let 'em come along my way and I'll give 'em 20s. an acre to plough my land ten inches; yes, and thank 'em besides', had not, however, been universally maintained. Steam work had been a most difficult thing to introduce and establish, both for the disappointed promoters, as well as for those like Fowlers, Howards and Smith who seemed to have more or less succeeded in finding a market for their engines and implements. However, while many users were now quite ready to admit steam power enabled work to be done which horses could not do, and admitted it was as cheap, they were often irritated by the frequent and sometimes expensive breakdowns with the machinery. Arguments on the pros and cons of the case were bandied between gentlemen of famous families, between farmers and the makers, between pro-steam and anti-steam farmers in the same villages, as well as among the labourers as they drank their Saturday night ale in the 'Horse and Harrow', 'The Plough' or 'Marquis of Granby'. The great question was whether the steam plough was a gimmick about to become a flop, or whether it was another instance of what blood had built, wealth would enjoy?

Lord Berners of Keythorpe, Leics., said in 1863 that he had found the steam plough could 'burst up' sun-baked land no horses could touch; a reasonably acceptable statement because it was summer work of this nature for which steam was most valued during the next sixty years. At the same time, up and down the country, it could be seen, how many gentleman farmers had cut down hedgerows, felled free-standing trees, and filled in ditches, in order to give the steam tackle elbow room. Wells or ponds were dug specially for the purpose of making boiler water available on the spot and to avoid the necessity of having to cart it long distances. Additionally, on the really businesslike farms where roundabout tackle was in use, stout oak or larch posts had been fixed to act as permanent tie poles for rope anchors. Digging holes for portable anchors proved an awfully irksome task and well nigh impossible on shallow soils overlying hard chalk.

Smith of Woolstone, promoter of the Smith's System, had a bit of a setback in 1863, because after loudly proclaiming how much better his crops were after steam tillage, the local Parochial Assessment Committee promptly put his assessment up by 10/– an acre. This extra charge naturally swallowed up some of the advantage of steam power. Another steam fan farmer claimed the size of worms had increased after steam cultivation with a corresponding improvement in the yield of corn.

On the other hand, there was much background murmuring from a fairly large body of disgruntled owners of steam tackle, who said the overall expense was high and breakdowns a too common occurrence. Already some machinery had been pushed aside in out-of-the-way corners to rust. It was, then, touch and go whether or not it was advisable for anybody to make further investment in the steam plough.

Favourable and unfavourable comments of a similar nature had already been heard at the lectures and debates of the R.A.S.E., where some responsibility rested to sift the evidence and come to an unbiased conclusion. The Society willingly accepted the challenge and set about this rather formidable task by appointing, in 1867, a Committee to 'Investigate the Present State of Steam Cultivation'. England was divided into three areas, to each of which a small sub-committee was sent on a personal tour of inspection and fact finding. In all, these three sub-committees visited no less than 311 farms where steam power was in use. At each place they either inspected the outfits at work, or asked opinions of the owners. During their rounds they saw Hornsby, Howard, Fowler, Ransome, Clayton and Shuttleworth, Kitson & Hewitson, Burrell, Barrows & Stewart, Tasker, Aveling & Porter, Ruston, Richardson & Darley (Kirton in Lindsey, Lincs.), Robey, Savory, Hayes (Stony Stratford), Horsfield (Leeds), Coleman & Morton and Smith (Coven, Staffs) engines engaged on land tillage work.

The Kent, Sussex, Hampshire and East Anglia party noticed a few portable engines had been fitted with reversing gear so that when working with some windlasses, a reversal of rotation on the engine effected the return pull on the implement. In parts of Sussex the autumn ploughing involved burying a longish surface mat of stubble, as the tenants were under obligation to their landlords to reap with the sickle in order to leave sufficient cover for such game birds as partridges and pheasants. Methods for hire, or costs given for the varied tillage operations differed from farm to farm, but a ten-hour working day was average for the counties included in this sub-committee's review. At Kimbolton, Northants, the Duke of Manchester had the unusual arrangement that he let out the work to the engine driver, ploughman and windlassman of his gang at 3/– per acre first time over, and 2/– the second. His Grace stood all working expenses. Upon enquiry at a farm at Scole in Norfolk whether the engine and implements were hired out, the owner said it was not done since he feared misuse of the equipment, or that the men would dawdle when away from their master's eye.

Mr Kersey Cooper, Agent for the Duke of Grafton at Bardwell, Bury St Edmunds, told how he had started in 1860 with an old set of Smith's tackle, obtained in exchange for a cart colt. A two-engine set was now installed, enabling one engine to help her sister out of wet places and up hills. When working on hillsides over a surface of big hard clods, it was the practice to rope the water cart up to the top engine to avoid overstraining the horses. An extraordinary discovery at this place was to find that the agent had converted the centre room of a labourer's type cottage into a sort of training and recreational centre. Besides being heated and lighted this room was well stocked with papers, books and games; while upon the walls were hung diagrams of machines used with steam power on farms. Coffee was provided and the room was open during the evenings to farm labourers and others for a small charge. That seems to have been a very early and commendable piece of staff training and welfare work, especially noteworthy as it was instigated by a member of the old British aristocracy, who as a class in general, have often been misrepresented as harsh and unjust employers of labour. Another innovation seen at Bardwell was an 'Engineer's House' on four wheels, with a stove, and living facilities for three men and two boys when the set was away from home on contract work in scantily populated districts where accommodation was scarce. The conclusion of this sub-committee was that steam ploughing was a success on heavy clay land, but doubtful on light soils. They had noticed a growing preference for two-engine sets.

One of the things observed by the sub-committee which visited the North Midlands and West was the unusual Chandler & Oliver (Hatfield) roundabout tackle powered by a Robey portable, on the hind axle of which was slung the rope windlass. They were told, however, that it had proved too much for one man to manage the windlass and look after the engine at the same time. On an average it was reckoned the top capacity of any roundabout tackle powered by a portable engine working 10 hours a day was 6 acres if ploughing, or 10 acres if cultivating.

Another owner mentioned the trouble caused by inexperience, when new wire ropes were supplied and attempts were made to fit them. His men had lifted the coils off one by one without taking the precaution of rotating the coil itself, with the result that a kink formed at every turn. A dodge now adopted was to lay a cart wheel upon the ground, drive a stout stake through the axle hole, lay the coil of new rope over the stake until it rested upon the wheel, and then with another wheel on top of the coil, pull the rope off evenly and neatly, with the coil turning between the two wheels.

Away west in Worcestershire, where already the famous Pershore plums were grown largely for making 'apricot jelly', members of this Committee found Benjamin Bomford on his model farm of strong tenacious clay at Pitchill. Ten of his fields had already been made into one of 120 acres, in order to give full steam ahead to his twin set of 1864 chain driven 12 h.p. Savory engines with return flue boilers, then at work with a Fowler-made cultivator. These two engines burnt 24 cwt of coal, costing 15/– a ton, between them during a day's work.

When the committee came to the farm of Mr J. Allin Williams (the man who had made the long drive home from the 1857 Salisbury Show), they found his 8 h.p. engine busy at work almost under the shadow of King Alfred's Hill, or as they termed it locally, 'The Stronghold of the White Horse'. This is the still famous ancient monument near Uffington in Wiltshire. By all accounts Williams was outstandingly steam-minded, for he had turned his malthouse into a foundry where they even 'melted iron', and he himself worked at a bench. In order to level off the fields on his farms to make work easier with steam tackle, no less than seventeen chalk pits had been filled in.

Other bits of incidental information picked up included the reckoning that it cost £46 a year to keep a horse and £15 for an ox. Several instances had been noticed where, on wet land farms, the steam machinery had proved too heavy and oxen had already been brought back into use again. The use of cultivators, or grubbers as they were often called, had become very popular. As previously mentioned, any attempt to break up hard dry clay land in summer was pretty well beyond the capabilities of horses. Indeed, one sad fact which came to light was that a certain farmer in a West Midlands clay land area had actually killed no less than ten horses by such overwork in 1866.

Contract cultivation work was noticed to be coming into greater favour, largely because it avoided any risk of capital outlay, either on tackle, or for the provision of repair facilities. With contract work, all the farmer had to do was provide coal and water, and then pay the agreed price for the work done, usually about 14/- per acre for ploughing, and £1 for cultivating twice over. The Herefordshire Steam Cultivating Co. and the Whitchurch Steam Cultivation Co. were two contractors which the committee saw well established in business. Foremen with the latter's sets were paid 1/- a week more than the men; not much, when one considers the post amounted almost to that of a travelling manager. From the evidence available, contractors were operating at a profit.

There was some antagonism against steam on the land and, as elsewhere in the country, there were instances where steam ploughing tackle had been abandoned to rust away. All in all, however, this sub-committee considered steam work was cheaper than if horses or oxen were used.

Much the same wealth of detailed information, including varying charges for work and wages paid, was gathered by the three members who went north over Yorkshire, Durham and Nottinghamshire. A small, but nevertheless interesting, feature observed up there was that on one ploughing engine pipes had been fitted to the waste water cocks on the cylinders, in order to conduct the escaping steam and water down to the ground. Many early engines, if not all, simply discharged the steam at the cylinder cocks, a noisy and disconcerting performance. As a result of this modification, it had been found that the driver could not only see better, but also horses were not so frightened. Near Carlisle a steam plough engine named 'Cain' was found filling in some idle time performing the thankless task of pumping pig manure. At that time coal cost 3/– to 4/– a ton in Northumberland.

On the staff side, one gang in Yorkshire was seen dressed in white duck suits, caps and jackets—regular engineer fashion. An objection to two engine sets heard in Durham was having to contend with the independent tempers of drivers who never agreed; one was bad enough, but two!

From what was reported by these all-England fact-finding teams, it is obvious that by 1867 a great deal of careful attention had been given to the steam engine so that it might have a fair chance to succeed. The well-wishers seem to have outnumbered the obstructionists.

Even though the R.A.S.E. Committees did not report enthusiastically on steam cultivation, on the whole they reckoned that within bounds, steam work was an economic proposition. The two really important observations were, of course, that tillage by cultivator implement during summer was much favoured; and secondly, that contractors had established themselves on a paying basis. Future steam plough work was to be developed on these two lines.

The complaints which had been made about breakages of materials were soon rectified by the manufacturers substituting steel for the former cast or wrought iron parts on both engines and implements. David Greig was in charge of the workshops at Leeds, and it is said that during his time, whenever any fitting broke due to weakness, all replacements were doubled in strength. Fowlers also made great improvements on drum and rope gears. One important lesson which had been learned was that in trying to overcome the burdensome weight of steam engines, strength of materials had often been subordinated to lightness in design. The Fowler plough engine of the 1858 Roo Dee Show was an elegant piece of design work; but just like the natty Savory engines, she would not have been strong enough to withstand the everyday buffeting of ploughing, or cultivating, on hard dry clays. It had been this rough work which had killed the farmers' horses, so it was only to be expected that these jobs, the worst of all, were given to the steam plough tackle.

When the Franco–Prussian War broke out in 1870 it had the

immediate effect of sending the two countries to buy wheat in the British market. Corn prices leapt up, and the demands from these continental buyers could scarcely be met. This state of affairs brought boom conditions to British agriculture, and the farmers were prepared to pay well for all the help they could get from the 900 odd steam cultivating tackles of various kinds then in commission. As may be well imagined, there was a pleasant air of prosperity over the whole field of British agricultural activity. The fact there was plenty of money about, meant buyers were well able to afford the £1,600 Fowlers were asking for a twin set of 2-cylinder engines complete. As a result, from about that time double engine tackles became increasingly popular. After the war was over, and the continental demands for our corn dwindled away, a terrible depression settled like a black cloud over British farms. One of the first effects, of course, was a great curtailment in the hiring of steam ploughs. On top of this misfortune there came another crippling blow in the form of an outcry that, on certain lands, the deeper tillage of the steam outfits had brought lower and sterile layers of soil to the surface, and so ruined the fertility of many fields. To a large extent this was perfectly true, because although on certain land the deeper ploughing had been of great benefit, on others the results had been disastrous, particularly on shallow soils overlying limestone or chalk. This mischance put steam work into ill repute, so much so in fact, that all through the 'eighties it was extremely difficult for contractors to get any work at all. Some share of the blame rested with the contractors who, finding that it was not possible to exceed $2\frac{1}{2}$ m.p.h. when the balance ploughs were set about the usual six inches, went deeper in order to speed up the work. As may be expected, orders for new engines fell alarmingly; and from 1881 onwards to the end of the century, hardly any steam ploughing tackles were sold for home use. The manufacturers were kept almost solely in business by the orders they received from overseas. At home, a good deal of old tackle was simply abandoned as scrap. It must be borne in mind however that the engines and implements made during the boom years 1870–74 had been put together on the good old British standard of 'made to last fifty years', and consequently once the initial demand had been met, a period of some stagnation would in any event have followed. It was not until about 1910 that the home steam plough trade revived.

This brings us to the end of the century so far as the general application of steam cultivation is concerned. The overall picture is one mainly of two-engine sets, owned by contractors, whose crews lived in a travelling van, moving from farm to farm during spring, summer and autumn. Although harrows were sometimes used, ploughing and cultivating were the principal tasks, with the emphasis on cultivating. The steam engine had not displaced the horse by taking on the tossing about of summer fallows, as well as cultivating or ploughing the stubble lands after harvest, but the number was slightly reduced. When the steam plough was first introduced it is said to have brought about an increase in the number of horses used to cope with the greater activity in agriculture generally as more land came under cultivation. By 1900 the position seems to have been that the engines came round to help out with the heavier work, which they did very well. Old farm hands said the steam plough was a 'Good old helper forward', and they also maintained the best crops of wheat were grown after land had been worked by steam.

5. McLaren 20 n.h.p. compound plough engine as shipped to
Australia in 1890

[Photo: Patent Office Library

View of Eddison's yard at Dorchester, Dorset, about 1885 during
the time Mr. John Allen was a partner

6. Aveling and Porter
single engine steam
ploughing tackle, 18

1907, Aveling and P
compound type plou
engine No. 283

1911 Aveling and Po
30 n.h.p. compound
steam plough engine
(Works No. 7380). T
indicated horse powe
rating was 140 and t
engine weighed 31 to
13 cwts., 2 qrs.

[Reproduced by court
Aveling-Barford, Gra

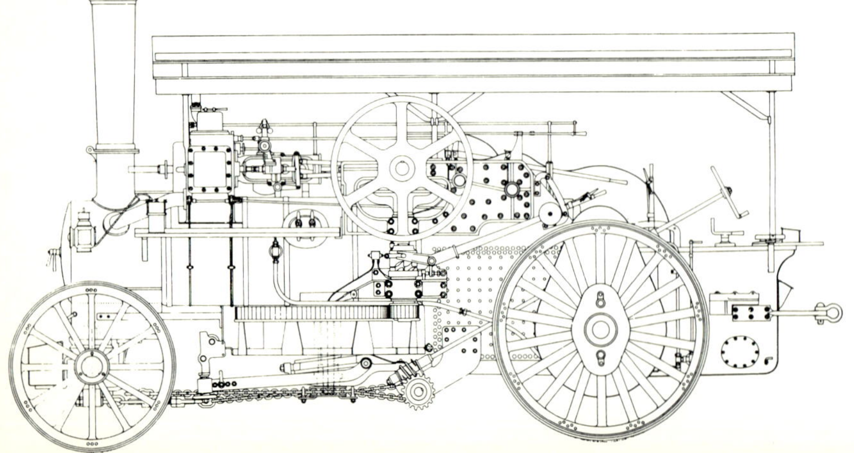

IX

Steam Plough Firms and the Engines they made, 1860–1900

One of the most interesting features in the early history of steam ploughing engines is the almost infinite variety of types which appeared both before and after 1860. So many private individuals or firms made attempts at constructing what they thought would be the ideal and trouble-free engine that there resulted an amazingly wide range of types, good and bad. It was only to be expected that in each design there were some unsatisfactory features, such as ropes coiling badly, engine not steaming freely, crankshafts prone to break, heavy coal or water consumption, a tendency to get stuck in soft ground, too heavy construction, too expensive to run, and so on. Now it was the inventor improver or manufacturer who took up the challenge thinking that if he could put this or that fault right by a different approach to the problem, all would be well and his 'improved' plough engine would be welcomed as a complete success. Many were spurred on by this thought.

Some idea of the individuality of outlook which led to so many fascinating types of engines appearing after 1860 will be gathered from the following review of the principal builders of plough engines in England.

J. & F. Howard Ltd, Britannia Works, Bedford

This company has been referred to previously as one of the pioneering concerns. It was very early in the field with, at one time, prospects of becoming a serious rival to Fowlers. However, Howards kept to odd unorthodox styles of engines. They never put the rope drum under the boiler in its proper place, nor did they build big sturdy engines for double work, so in consequence the firm failed to get into really big business as steam plough engine

E   65

makers. But what they did is of considerable interest because it showed marked originality of outlook.

By 1862 there had appeared a 10 h.p. self-moving Howard engine intended for roundabout cultivation. The firm's ploughing machinery was mostly of the roundabout pattern and was developed from the tackle they first made in conjunction with William Smith of Woolstone. We find Smith had written to the 'Live Letter Column' of the December issue of *The Mechanics Magazine* in 1862 complaining about the editor's earlier remark that, 'Messrs Howards had improved Smith's system out of all recognition'. To this Smith had replied that he still had 188 customers and that he could even plough and sow in one operation.

On April 7, 1863, patent No. 880 was taken out by James Howard, Edward Fenny Bousfield and John Pinney for a most peculiar plough engine, which was particularly odd because it had a boiler divided into two halves connected by a common overhead flue. This flue held the two sides of the boiler together almost like the arch of a bridge, while the covered space or tunnel underneath was used to house the inverted engine. Another point of interest about this boiler is that it was constructed on the water tube principle, where the water is contained inside the tubes, instead of outside them as in ordinary multitubular boilers. In marine and factory installations, boilers of this design had already proved successful as quick steamers, so there was good reason to try one out on a ploughing engine. Unfortunately, the experiment did not turn out well, any more than a similar trial made sixty-five years later by Sir Nigel Gresley with a Yarrow water tube boiler on a high pressure express passenger locomotive of the London & North Eastern Railway. Howards claimed that since their boiler was hung transverse or astraddle the direction of the engine's movements, there was no risk of the water in the boiler running to the front end and so uncovering the firebox top when running downhill, either in the field or on the road. These particular engines were also three-wheelers, with one wheel in front and two behind. This practice was seldom followed in this country, though it was fairly common later with North American plough engine builders. When the Market Drayton Steam Ploughing Co. bought a pair of these unusual Howard ploughers in 1865, they found the boilers leaked badly, the teeth of the gear wheels broke easily, and in general it was considered that the engines had been most trouble-some, and a disappointment to the owners.

Other extraordinary engines followed from the Bedford Works, for instance a Fairlie-boilered model with one central firebox serving a tubed boiler section on either side. No one could say that Howards were not doing their best to find a boiler that might prove better than the multitubular type most commonly in use. Regarding the unusual appearance of the firm's engines generally; a rustic was heard to remark in 1871, 'A bad 'un to look at is often a good 'un to go'. No doubt this was a well-meaning remark, not strictly correct, because the opposite was the case in this instance and also in many others where engines lacked looks as well as ability. The doubtful performance of Howard's early engines was confirmed in the report of some trials held in 1871, when it was said they were the heaviest on coal consumption, and their gear teeth stripped off wholesale. Oddly enough, it was mentioned in a lecture on steam cultivation given in 1867 that the method for testing Howard's engines was to run them at speed over logs of wood laid on the ground. One would have thought such drastic treatment ought to have proved the gear teeth before delivery to a customer.

Having been so obliging as to spend time and money proving for the benefit of others that neither water tube nor Fairlie boilers were really suitable for land work, Howards turned out in 1876 what was probably their most successful plough engine known as the 'Farmer's Friend'. At eleven tons weight this engine was quite big, and designed either for double engine or roundabout work as required. Even if the firm bowed to custom by installing a locomotive or multitubular type boiler, prettily capped with a dome, they still refused to be copycats by locating the rope drums on a horizontal shaft at the rear of the engine instead of under the boiler. The single cylinder engine with its flywheel and other gearing was placed alongside the driver on the manstand where, they said, it was under his eye. It was also simple in this position to couple the engine to the drum drive. The rope was led forward under the engine, round a large free-running pulley below the smokebox, and then out at right-angles to the implement. The maker's claim was that with this method of hauling the rope the engine did not slew round at the front, as often happened where the tackle had boiler-slung drums. Also, as the engine was not mounted on the boiler, the boiler was not subjected to motion-work strains in addition to those of internal steam pressure. These engines had a fairly wide ciruculation and some of them were still at work in Britain after the First World War.

One 'Farmer's Friend' engine was sold off the firm's stand at the 1877 Liverpool Show to Mr Robert Stephenson of Burwell, Cambs., and it has a story worth mentioning. It continued to work at Burwell until 1928, when it was bought back by Howards for sentimental reasons and because they wanted to preserve it. A picture in the *Daily Chronicle* of March 18, 1929, showed the engine proceeding under its own steam on a four-day journey by road from Burwell to Bedford captioned, 'The Return of the Native'. According to contemporary report horses shied at it on the way. Unfortunately, Howards, like Burrells, Aveling & Porter, Garretts and others, became members of the ill-fated Agricultural & General Engineers Ltd, formed shortly after World War I, in an attempt to ride out the years of depression which had settled on the engineering industry. When in 1932 this confederation failed Howards, like Burrells, were completely ruined. The Bedford Works passed into other hands who cannot be blamed if they did not happen to be steam engine fans. The actual end of the old engine is uncertain, but it appears that when the space upon which she stood in the old works yard was required for expanding business some years later, she was cut up for scrap and that was the end of the last known example of this type of plougher in Great Britain.

One of these 'Farmer's Friend' engines was sold secondhand by Baldwin Bros. of Wadhurst, Sussex, to the Great Western Railway, and used for a time on cable shunting duties at a South Wales Colliery.

Aveling & Porter, Rochester, Kent

As a firm, Aveling & Porter of Rochester were perhaps best known for their steam rollers. They did however build a considerable number of engines for ploughing between the years 1863–1921. Old Thomas Aveling, the founder of the firm, was born in Suffolk in 1824. When his father died at an early age and his mother remarried, he went to live at Rochester in the new home of his clergyman stepfather who by all accounts was a bit of a hard-handed character for a parson. While apprenticed as a young man to a farmer near Canterbury, Tom gained a knowledge of agriculture implements and from this he turned to engineering starting, in partnership with Mr Porter, a steam engine and implement works on the west bank of the Medway at Rochester. Just when he first began to show an interest in steam for land work is not certain, but he did once say, in 1875, while giving evidence at the Sittingbourne petty sessions during a police prosecution for alleged highway smoke nuisance concerning a double engine set of Aveling & Porter ploughing engines, that he had first experimented with steam cultivation in 1852. Certainly Aveling bought an engine and roundabout tackle from John Fowler at Leeds in 1859, presumably in order to formulate some ideas of his own for future construction work at Rochester. Later on the two firms co-operated in patents connected with steam engines. One set of engines from the Kentish works went out to the 1869 ploughing trials held at Beauvais in France. There is also a story of how on one occasion the Emperor of France, Napoleon III, was amazed when Thomas Aveling showed off a bit by actually ploughing up a macadamized road. Whether or not the Emperor bought any Aveling & Porter engines I do not know, but he did acquire a set of Fowler machinery.

Following closely upon Fowler's designs, Aveling's engines were usually normal looking with their rope drums under the boilers. The power sizes varied, but generally speaking Aveling & Porter plough engines were on smaller lines than Fowlers' who placed the emphasis more often than not on massive engines. Their 1874 engine was for twin engine work and is described in the firm's catalogue as having a boiler made of best best steel, a firebox of Lowmoor iron; was rated at 6 h.p. and sold with 400 yards of steel wire rope coiled on its drum ready for work. The smokebox, like those of most engines at the time, appears too short for good even steaming. Within bounds, the greater the volume inside the smokebox, the more evenly the blast vacuum is absorbed, so giving a steadier pull on the fire. Like Savage, his contemporary up at King's Lynn, Aveling gave an all-purpose touch to one type of engine by making provision for the easy removal of the rope drum. There were also facilities similar to those on Burrell or Fowler 6 and 8 h.p. types for setting back the front wheels so that the ploughing engine could be converted to a normal traction for haulage work. Since neither of these operations could be termed handy or be quickly accomplished, it seems unlikely the engines appealed to many would-be owners. While mentioning the setting back of front wheels on plough engines it will probably have been noticed from the illustrations these wheels are usually set well forward. In fact, before the coming of longer boilers with extended smokeboxes after about 1900, a special bracket protruded forward from under the smokeboxes, for the express purpose of setting the front axle well out in front. Even if this arrangement somewhat spoiled the appearance of the engines, it was essential to give the working rope a wide angle forward, so necessary when finishing up against headland or hedgerow on the slant and where the pull might easily be at about 45 degrees between the two engines.

Pulling the implement up against the hedge bottom when finishing, or getting into corners or 'gorings' as they were called, also demanded good clearance for the rope. Even so, the rope frequently did bear against the front wheel, with the result that the hard steel strands sawed into the relatively soft mild steel rims. Like tomcats' ears, most ploughing engine front wheel rims were 'battle scarred'.

The firm also experimented during the 'seventies with right-and left-handed ploughers, that is to say the flywheel and the upright shaft driving the rope drum were placed on the 'field' or implement side of each engine to ensure that the strain of pulling the rope was taken on the driven side. By this Aveling calculated less stress was imposed on the stud or stub axle upon which the drum was suspended from below the boiler. However, whatever the advantage of this arrangement it never became standard. According to Mr E. E. Kimbell, he found plough engines built by Aveling & Porter were very noisy in their drum driving bevel wheels.

Quite naturally, Kent presented a ready market for Aveling & Porter's engines, whilst others were hoisted aboard ships to take long journeys and spend their working lives in South Africa, Australia, France, Greece or Egypt. By and large, however, the firm's cultivating machinery construction always seems to have taken second place to the making of traction engines, steam rollers or steam wagons. All the same, they were among the top four British steam plough builders.

Aveling & Porter's trade emblem was the well-known rampant Horse of Kent over an 'INVICTA' scroll, and closely resembles the coat of arms of the County of Kent. The rearing horse motive is said to have come over on the war banners of Hengist and Horsa, raiding Danish chieftains who arrived as armed settlers on the Thanet coast in 449 shortly after the Romans had evacuated Britain in A.D. 400. The word Invicta (meaning in Latin unconquered), was later added to the county arms in the following circumstances. When William, Duke of Normandy, had beaten King Harold in 1066 at Senlac, he first made a successful march on London intending to press north and subdue the rest of the country. But the castle of Dover held out, and he did not dare to move away while this threat of being harassed in the rear remained. In face of this difficulty he was inclined to compromise, so in 1067 he arranged to meet the nobles and high clergy of Kent at a place near Springhead Crossroads, Swancombe, on the Canterbury Road. On William's assurance that he would not ravage the castle of Dover, the Men of Kent accepted him as king and promised not to resist him further. He on his part allowed them, as a special concession, to retain their ancient Saxon privileges. The outcome of all this enabled the Men of Kent to consider themselves unconquered, and so we have the motto. The lovely Aveling & Porter rampant horse and scroll in burnished brass has carried these two incidents from the history of Kent far and wide about the world for all to admire. It is without doubt the most attractive trade mark of any British traction engine manufacturer. A small sized brass horse and scroll was fixed on the smokeboxes of all their plough engines. Strictly speaking of course Aveling & Porter's Works were in the domicile of the Kentish Men, on the west bank of the Medway.

A carved and inscribed portland stone memorial, surmounted with the county crest, stands at Springhead Crossroads, Swancombe on the main A2 London road near to the legendary meeting-place of William the Conqueror with the Saxons of Kent.

Tuxford & Sons, Boston

It is an odd thing that the widespread arable county of Lincolnshire, famous for the amount of steam cultivation which used to be carried out over its broad acres, cannot claim much in the way of cable plough engine manufacturers. However, Tuxford & Sons of Skirbeck Iron Works, Boston, did make a few traction engines which were employed on roundabout ploughing using manilla ropes; and so it is not out of place to include one of their engines.

Tuxfords were pioneers in the steam engine business, for they had certainly sold a portable farm engine by 1842. Their traction engines were strange and individualistic in design like 'Progress' the 'Improved' 10 h.p. engine used for many years on the Thorold estate at Syston, Lincs. Her chief features were a single cylinder over the firebox, a chain drive through the differential-fitted second motion shaft, a water tank combined with smokebox to warm the feed water with waste heat, volute springs incorporating rubber pads on all four wheels, a hand brake acting on one front wheel and a boiler with a working pressure of 120 lb. She was also peculiar in that she ran firebox first.

'Progress' was bought in August 1872 for £420 by Sir John Henry Thorold, Bt, of Syston Hall, Grantham, who intended to use her for general haulage and belt work, as well as for Fisken style roundabout steam cultivation on his largish estate at Syston. The old estate accounts show that the three men who brought the engine from Boston were provided with refreshments at a cost of 3/– the lot. It may well be that the engine was actually driven from the Skirbeck Works to Syston under her own steam, accompanied by Palmer, the man who had been selected as her future driver. The cultivation tackle comprised of a secondhand set of Howard's steam harrows bought from Amies & Barford, of Peterborough for £15; while a Fisken four-furrow balance plough for £80, seven-tine turning cultivator for £70, three hundred yards of $\frac{3}{4}$ inch manilla rope for £6, together with various other sundries and spares for Fisken style work, came from the Ravensthorpe Engineering Co., Mirfield, Yorks. The entries in the old 'Engine Book' rather indicate 'Progress' gave a fair amount of mechanical trouble. In less than a year after her purchase, Hornsbys of Grantham had either repaired or replaced the connecting rod. In December 1873 it was necessary to return the engine to her makers at Boston for both driving wheels to be taken off and strengthened with angle iron and plain rings, as well as the fitting of a new volute spring. In January 1876 she was once again on her way back over the twenty miles to Boston, for attention to some defect not stated. During her absence this time Messrs Tuxford sent on loan for three months one of their 8 h.p. New Expansion Engines; she was consigned by the Great Northern Railway at a cost of £1 9s. 0d.

After 'Progress' returned home, there was some further trouble with her intermediate shaft, which went to the Works at Boston for repairs described later on the bill as 'Your intermediate shaft for traction engine cut and pieced with new iron and shaft turned up—£2. 4. 3d.'. Besides the initial 1872 outlay of £770 on the engine and cultivating equipment, another £800 was next spent on such barn machinery as corn mill, chaff cutters, cake breakers and pig feeders. Sir James Thorold told me in 1961 that he remembered that when the engine was not out and about on ploughing or running work, it was used to provide the steam to boil the pigs' dinners. Before they invested in steam themselves, the estate had been in the habit of employing Hy. Yates, Steam Cultivation Contractor of North Parade, Grantham. In May 1872 he cultivated fourteen acres near the railway siding at the cost of 22s. 6d. per acre.

My father often talked about this old engine. He well recollected hearing it 'barking away', as he termed it, while it was tearing up the heavy clay land alongside Jericho Woods near Barkston Station. This would be in the 1880's when my father was a young signalman in his box on the Great Northern Railway. He was careful to impress upon my brother Sidney and me how 'The engine ran backwards with the men in front and the chimney behind'. On one occasion, a few years before the turn of the century, it got out of control, and ran away while hauling two three-wheeled 5-ton trucks loaded with coal up the Poultry Yard Hill in Syston Park on its way to the private gasworks which then supplied the Top Hall with gas. One of the crew was killed, casting so much gloom on the venture that it was not long before it was decided 'Progress' should be taken off road and ploughing work. With her wheels removed, she was installed in a big wooden-sided and corrugated iron roofed shed in the estate woodyard and given the permanent job of turning the big circular saw. When her own boiler began to wear, a new and separate vertical one was placed beside her, and with steam from this source, she puffed on for many a year as the sawmill engine. Just after the 1914–18 war, I was taken by my father, who had more than a passing interest in steam, to see this engine and although on that particular Saturday morning the old veteran was not at work, I did have a chance to look at her as she loomed big and dark above me in her shed. When the boiler inspector came round in 1924 he found the substituted boiler had worn 'as thin as a sixpence' at the bottom. That was the end; 'Progress' had finished her fifty-three years' career on the Syston estate. Very soon afterwards she was cut up for scrap by Spick & Co. of Grantham. Had Tuxfords still been in business in 1925, they might well have been proud of her.

Savage Bros, St Nicholas Works, King's Lynn

Frederick Savage began his working life as a boy scaring crows on a Norfolk farm for 2s 6d. a week. He eventually founded, in the year 1850, a small agricultural engineering works in the back yard of a pub at King's Lynn. It was not, however, until 1872 that he built a traction type engine intended for land cultivation. This was a large chain-driven 10 h.p. type, which operated roundabout style work by ropes coiled round the rear wheels which were jacked up clear of the ground during ploughing duties. As his business brought him into daily contact with the not-too-well-off farmers of the Norfolk borders, he probably heard many opinions about what these small men wanted in the way of steam engines on their farms. The great expense of the two-engine sets was beyond the purse of all but the relatively wealthy. Another thing was that most engines at that time were not powerful enough when it came to trying to drive them out of any hole, rut or soft place into which their hind wheels had sunk, and, if a steam engine came on to the farm at all, why could it not be used for all the various jobs where power was needed, whether cultivation, haulage or belt work? With a view to meeting these needs, Fred set to work again and in 1876 turned out what he called his 'Agriculturalist', single cylinder type engine. Following the pattern of his 1872 prototype, this engine weighed thirteen tons, and was designed to perform haulage, belt work or roundabout style cultivation as and when required. It had the same arrangement of wire ropes coiled in special recesses around each hind wheel. By jacking up the rear of the engine, packing baulks of wood under the bunker and removing the curved cover plates around the wheels, the wire ropes were made ready for cultivation work. Then the hind wheels were turned slowly to uncoil both ropes. The left-hand rope was led forward and the right backward. They were then guided one on either side of the field, and coupled to the implement, one at the front and the other at the back. When the engine was driven under power with both the back wheels turning 'in the air' so to speak, the rope attached to the front of the plough was wound in to haul the implement across the field. By reversing the engine, the other rope was coiled and the plough pulled back again. This process continued progressively towards the engine, till the field was finished. A system of three-wheeled frames, known as rope porters, carrying free-running rope pulleys was used as support to hold the longish ropes used with roundabout sets and prevent them from rubbing on the ground. Incidentally, similar rope porters were also used with two-engine sets in the early days of steam ploughing. Savage also fitted a third and very low gear to his 'Agriculturist' all-purpose engines. This extra low gear helped considerably if an engine became bogged down. Engines so fitted could be really speeded up and 'have a go for it' when trying to lift themselves out of holes. Up to 1885, about fifty-five engines of this type were built at King's Lynn. Over half of them were sold on the Continent, and some were straw burners.

Charles Burrell & Co., Thetford, Norfolk

The famous old Norfolk firm of Charles Burrell was founded as an agricultural engineering concern as far back as 1770. They concentrated mainly on a prodigious output of various forms of traction engines, paying scant attention to the construction of cultivating engines.

Their 'Universal' 10 h.p. engine of 1880 was intended principally for single engine work and was constructed in accordance with patents originated by Everitt, Adams & Co. at Ryburgh, Norfolk. The rope-hauling drum on this type of engine was fixed on the-right hand side in an upright position, where it was considered handy for direct spur wheel drive from the crankshaft. Two speed ploughing gear was employed. The drum in this rather unusual position was an experiment and perhaps it made the engine look a bit lopsided and ungainly. This was another example of an attempt to provide a dual purpose engine, because it was claimed the rope drum could be easily removed when the engine was required for haulage work. In all, between 1880 and 1914, about forty of these side drum ploughers were made, mostly for Germany in the immediate pre-1914 era.

Just over a hundred ploughing engines, for use in pairs, with drums hung below the boiler, were also built spasmodically at Thetford from 1867 to 1914. Chain drive was a feature of the earlier engines. The illustration of a single cylinder geared Burrell plougher shows clearly the good appearance and outline so typical of Burrell engines.

The later type Burrell ploughing engines were double crank compounds with geared drive, usually built to special orders for foreign purchasers.

John Fowler & Co. Ltd, Leeds

A good deal has already been said about Fowlers but as they occupied so pre-eminent a place as steam plough builders and played the leading part in the development of this trade from the founder's death to the end of the nineteenth century, further mention of the firm in this section of the book is fully warranted.

By 1862 Fowlers were already thinking of setting up subsidiary workshops in Austria. In 1861 one of their clip drum ploughers was taken by R. W. Eddison to Philadelphia. A single engine roundabout set was exported to America before John Fowler died in 1864, but stood three years in New York without being used owing to the upset caused by the Civil War. Later on, about twenty-eight tackles were actually sold in the U.S.A., mainly double-engine outfits.

Experimentation, often in co-operation with inventors outside the company, was a constant feature of the work of the firm. In the year of his death in 1864, John Fowler had been so adventurous as to join forces with a Mr Webb of Uttoxeter. They proposed to burn oil as fuel in an engine and also to condense the steam; these proposals and investigations were both important. Many people have toyed hopefully with the idea that great advantages might be gained from substituting oil for coal as fuel in the fireboxes of steam road vehicles. Unfortunately, it has been only in marine work, in power stations, or on railways abroad where oil was cheap, or the lines ran through forest areas subject to fire hazards from sparks, that oil has ever gained any great hold over coal. For oil burning, special burners requiring air or steam pressure to vaporize the oil are necessary, while the sickly smell of the burning oil is usually objectionable to drivers. By comparison, a coal fire is an incredibly simple affair to manage. Condensing of steam offers two principal advantages: (i) The condensate in the form of hot water can be returned to the boiler thus saving heat energy. (ii) In the case of ploughing engines the use of the water over and over again, would reduce the endless and expensive business of carting water to the engines as they worked in the fields. Unfortunately, nothing of value to cultivating engines seems to have come from these experiments.

The first good traction engine type steam plough built by Fowlers left their works in 1861. In 1864, the firm began to change over from two cylinder engines to a single cylinder type. What prompted this change is not altogether clear, unless it was that twin cylinders ate up too much steam, or that it was less expensive to make only one cylinder and set of valve gear. Single cylinder engines have a bad habit of not being able to start when their one big end stops either on the front or back position of the crank. The piston is then at the end of the cylinder and the crank cannot be turned. On two-cylinder engines the cranks are set at 90 degrees, so only one of them can be dead centre at any one time. Incidentally if at a rally you should see a driver pulling his flywheel over by hand in order to get the engine started, that will be a single-cylinder engine playing its usual trick. Anyway, the single-cylinder engine began to take pride of place in the erecting shop at Leeds, and in 1869, the first of the famous 12 h.p. singles was put together.

Another old custom which was dropped by Fowlers from about 1868 onwards, was slant shaft drive between the crankshaft and the road wheels. As the name implies these shafts were arranged slant-wise from a bevel drive at the upper or crankshaft end to a second bevel meshed with an annular toothed wheel fixed to the hind axle. On the 14 h.p. plough engines so fitted, the left-hand rear wheel was driven by a pin and the right-hand one through a friction grip ring. Although it was quite a logical means of transferring the drive by one single rotating shaft, these shafts were inclined to break. The next choice was a train of toothed gear wheels, probably because they offered more scope for two road speeds. Already the lesson had been learned that insufficient leverage had been allowed in early gear arrangements. The gear ratios had been too high for the heavy engines to get about easily on soft land where the going was difficult; so the new Fowler-designed low speed gear gave more pulling power than previously. Gear wheels were, however, rather troublesome for some years to come because of the brittle metals used in the manufacture of their teeth, which were very easily ripped off, if the engines had to be 'knocked about' at all in a rough place. It was not until 1871 that Fowlers were able to offer for sale a traction type engine with all steel gearing, and so fulfil their founder's expressed view that 'full steam effort should not break gear teeth'.

As may be expected from any undertaking with a ready and open attitude to try anything once, Fowlers came out with some odd machines. One example was the 28-ton plougher shown at the 1869 Manchester Show, or the Allison's Patent vertical pot-boilered machine of 1870 which did not look in the least like a proper ploughing engine.

Construction of the 12/14 h.p. singles went on all though the boom years of the Franco–Prussian War. Many hundreds of these engines, varied with minor modifications from year to year, were actually built and sold both at home and abroad. In the British Isles they were widely scattered and proved well made, rugged and reliable engines capable of doing a hard day's work with little risk of breakdown. Some of these two-engine sets were actually still at work in the 1930's, although it must be admitted re-boilering and other extensive repair work had given them a 'second wind'. As a class they were not particularly economical, for it used to be said they needed a coal mine in front of them, and a river of water behind them, in the field where they were at work. Be that as it may, these engines could work and keep at work. They formed the backbone of the steam plough stud which slogged through the hard hot work of a good fifty British summers. To many of the old school of steam plough admirers, these engines were the be-all and end-all of steam engines on the farm. They were certainly the best known and probably the most famous of them all. My first introduction to cultivating engines came through them, because the Sleaford contractors who did all the work during the First World War in our village of Barkston, Lincs, always sent a set of singles.

When the general building of single cylinder plough engines terminated in Fowler's works in the year 1877, it was a matter of special significance because it ushered in the era of the compound engine, which for all practical purposes held its own until the end. A small number of single cylinder plough engines were built subsequently at intervals until the nineteen-twenties. A compound engine has two cylinders, one bigger than the other, so that the steam may be directed first through the small one and then into the larger as it continues to expand, and so give up its pressure right down to the last few pounds. The first and smaller cylinder is known as the high pressure one. Steam enters this cylinder at boiler pressure of, say, 180 lb. and pushes the piston one stroke (with cut-off at about 75%) and in so doing, its pressure falls to around 75 lb. It then exhausts into the low pressure side steam chest, passes into the second cylinder and in pushing the piston there another stroke gives up all but about 16 lb. of its pressure. The exhaust steam is released to the chimney at just above atmospheric pressure. This accounts for the low 'huff-huffing' sound heard at the chimney of a compound, as compared with the sharper, noisier barking of a single-cylinder type. Steam passing through a Fowler compound expands to eight times its original volume, and the cylinder sizes were so designed that the work obtained from each was equal. The first Fowler compound plough engine was turned out in 1881 when the firm claimed an economy of 40% in fuel and water. This claim was perhaps a little high for savings which undoubtedly existed to a lesser degree. From that beginning compounds were pretty well standard at Leeds for all future double-engine tackle. During the following twenty years, the majority of those built were sold around the world, rather than in Britain where the market was already saturated with the famous old singles which had been built to last a lifetime.

It would not be right, however, to think that from 1869 onwards all Fowler's ploughing tackles were two-engine sets. The cry of the small farmers that they could not afford the expensive double sets was heard also at Leeds, and to meet the small man's market, Fowlers built two designs of single-engine apparatus which they introduced just before 1914. In the first or tail rope layout, the engine had two drums, one above the other with the top drum carrying 300 yards of $\frac{5}{8}$-inch rope by which the implement was pulled on the inwards or working run only. A light 700 yards $\frac{5}{16}$-inch rope on the lower drum was used to haul the empty implement back to the opposite anchor. The second system also used double drums and although two-way cultivation was performed through two ropes of equal strength, the one to pull the implement outwards was 900 yards long, as it had to reach over the double length between engine and automatic self-moving anchor opposite. Reading Corporation Sewage Farm used a double drum Fowler engine from 1903 until after 1926. Other similar tackles went out to France and Algeria where they were used to break up hard vineyard soils 24 inches deep at a rate of around 2½ acres a day.

By 1896 Fowlers had built, for trial purposes, twin-engine sets with the rope drums slung vertically on the sides of the boilers, and driven directly from the crank shaft through spur gears. As with similar types made by other manufacturers, these engines looked ungainly and definitely unbalanced. The rope was led backwards under the ashpan to a free-running sheave wheel at the rear of the bunker, from whence it turned away at right-angles to the implement in the field. Quite probably, as was found with the Burrell-Everitt engines, rope coiling was not too good under this arrangement. Very few Fowler engines of this kind were ever built, and so far as land cultivation was concerned, the honours rested with the drum-under-the-boiler style of engine.

In conclusion, I would like to mention that until recently we had with us a personal link with the steam plough engine building days of the great Fowler firm. Alf Pepper, who was for many years their Chief Engine Tester and latterly Chairman of the Fowler Veterans' Association at Leeds, died 3rd October 1964, aged 82. He was to the last an inveterate Fowler fan and never missed an opportunity to extol the virtues of the Fowler engines against all comers.

J. & H. McLaren, Ltd, Leeds

McLarens came into the ploughing engine business in 1876. They were thus latecomers rather than pioneers, yet they soon succeeded in making a name for themselves. When Fowlers finally gave up their former great part in the trade, McLarens picked up the flickering torch and carried on for a while supplying spares for Fowler cultivating tackles.

The founder, John McLaren (later Sir John), was born on April 1, 1850, of Scottish parentage at Hylton Castle, County Durham, where his father farmed over a considerable acreage. John served an engineering apprenticeship, much of it on the shop floor as was customary in those days, with Black, Hawthorn & Co. of Gateshead-on-Tyne who were engaged in the building of locomotives, marine and stationary engines and boilers. When his training was completed in 1871, John joined the Ravensthorpe Engineering Co., Ravensthorpe, Yorkshire. As we know, this firm

7. Howard 'Farmer's Friend' single engine ploughing tackle moving
to next job in East Anglia (about 1920)

Burrell 16 n.h.p. single-cylinder plough engine No. 850.
Built 1881

[*Photo:* Major R. J. W. Ind collection

8. Darby steam digger working at Peldon, Essex, July 11, 1904.
Front view of machine. The digging forks are hidden
at the rear

[Reproduced by courtesy of the Road Locomotive Society

Wallis and Steevens steam tractor with plough bodies attached
either end, 1909. (In this view the engine is moving backwards)

[Reproduced by courtesy of Wallis and Steevens

had earlier taken over the rights to make Fisken type implements for roundabout work, so the young man doubtless found himself right in the middle of the manufacture of cultivating machinery intended for use with steam engines. In view of his home farming background, it needs little imagination to assume that he found the job to his liking. In fact, by sheer hard work and engineering ability, he soon rose to be general manager of the Ravensthorpe firm. In 1876 he left to join with his brother Henry in founding their own steam engine and implement factory, known as the Midland Engine Works, at Leeds. Why they chose Leeds must have been for some reason best known to themselves, but here they were in the city where Fowler's great works were already established as the main centre in the world for manufacture of steam cultivating tackle. By all accounts, however, the two firms got along well together in a very neighbourly fashion. Neither tried to encroach upon the other's market. Many years later, when McLarens had secured the Egyptian trade for steam tackle, Fowlers gracefully conceded that territory to them.

In appearance and general design the steam plough engines from the Midland Engine Works were very similar to Fowlers, and like them, McLarens preferred the heavy two-engine sets. Apart from the thinner chimneys with McLaren style caps, upright shaft drive to drum through toothed pinions instead of the heavy-jawed dog clutch, the valves positioned on either side of the cylinders rather than over them, and a straight-spoked flywheel, there was not a lot of difference in appearance between the two makes of engines. A specialist in mechanical detail would, of course, notice that McLaren ploughers had right-hand steering, just the opposite to Fowler's engines.

How far the firm had gone in plough engine design by 1890 is well indicated by the illustration on plate 5 (p. 64) which shows one of their engines due for shipment to the raisin farms of Messrs Chaffy Bros at Mildura, Australia. The boiler was made of Siemen Martin steel, there was a large water tank holding 400 gallons, and the large open grill above the bunker marked the engine as a wood burner.

Just as the Fisken enterprise had had a religious background, so also was it the case with the McLarens. Sir John was a devout and prominent Congregationalist, active as a Sunday School Superintendent and chapel worker all his life. His father used to tell proudly the story that he was the first to demonstrate the reaping machine in his own district up north. On his part the son could say that during the seventy years of his life he attended no less than fifty-two Royal Shows. He received his knighthood in recognition of the outstanding service he gave to the organization of munitions manufacture in Leeds during World War I.

X

Steam and the Steam Plough

The mobile steam engine is fast becoming a thing of the past, as it is swept away under the great bow wave of the popular and more economical diesel engine. It may well be that soon even the elementary principles of steam power will not be readily recalled. So I will give here, in simple terms, a brief description of the principles and general construction of a steam plough engine.

Apart from wind or water power, man's first successful attempt at getting work done by power other than muscle was burning combustible substances and using the heat so released to turn water into steam. The expansive properties of this steam gas were then utilized. Boilers of various kinds had, of course, been in use since the days of the ancient Egyptians, but after George Stephenson's time a boiler which he developed, and known as a multitubular or locomotive type, had been in common use. The chief feature of this type of boiler was a cylindrical shell through which passed a number of smoke tubes sealed back and front in tubeplates. Attached to the hind end and forming an integral part of the boiler, was a water-enclosed firebox. From the fire on the grate, heat radiated to the water around the firebox, as well as to the water surrounding the tubes conducting the hot gases to the smokebox. By this simple arrangement about 90% of the heat obtained from the fuel was transferred to the water for steam generation. In the case of steam ploughs, the boiler with its firebox and smokebox form the essential framework. Together they provide the main support for such fittings as cylinders and motion which are on top and gear and road wheel bearings which are fixed at the side. From below the smokebox a bracket extended to support the axle of the front wheels. Unlike the railway engine, where the boiler is lowered into the frames, steam ploughs were built around the boiler unit.

In caring for the boiler when at work, the chief concern of the steam plough driver was to see that water was kept above the top or

83

crcwn plate of the firebox at all times. If, however, the level falls below the crown, the steel plates are in danger of getting red hot, and becoming soft, like iron heated by the blacksmith prior to bending. The result at the best is distortion of the crown plate, or most likely its sudden collapse inwards under the pressure of steam on top—in other words a violent boiler explosion. A simple automatic device is provided in the form of a lead or fusible plug. This is a short, thick, threaded brass bolt with its $\frac{3}{8}$-inch centre filled with lead, and the whole assembly is screwed into the firebox crown from the under side. Immediately overheating begins, the lead melts to open up a hole through which the steam bursts and quickly extinguishes the fire. This is known to drivers as 'blowing the plug'. One or two glass tubes fixed to the outer rear of the firebox show the driver the level of water in the boiler. The bottom of the glass is positioned just above the level of the firebox crown, so it is a matter of never letting the water 'go out of the bottom nut' as they say.

Either an eccentric-driven force feed pump, or an injector under the control of the driver, supplies water to the boiler as and when required. An injector is a compact and cunning device comprising a series of variously shaped cones, through which water, as it mixes with a jet of steam, is first given speed, and then in the last or divergent cone of the injector, given pressure sufficient to deliver the water against the boiler pressure. When working correctly, injectors give out a distinctive soft satisfying singing note from the open end of the water overflow pipe. In the beginning, boiler pressures were in the region of 40–50 lb. sq. in., but eventually reached 200 lb. in later years, when good strong $\frac{7}{16}$ inch plate boilers became available. A twin spring-loaded safety valve, mounted on the top of the cylinder steam space casting, released excessive boiler pressures.

The power of the plough engine is derived from one or two cylinders bolted down on to the upper side of the boiler at the smokebox end. Connecting rods couple the pistons to the crankshaft, located over the top of the boiler or firebox at the driver's end. Steam distribution is usually effected by what has come to be known as Stephenson's link motion, invented in 1842 by a certain William Howe of Chesterfield then in the employ of George Stephenson on railway locomotive work. In recognition of his invention, a learned society presented Howe with two hundred sovereigns, while Stephenson at once used the gear with great advantage on the locomotives he was building. The Stephenson valve gear functions by having two eccentrics set upon the crankshaft: one at about 106 degrees in front of the crank for forward gear, and the other at the same angle behind the crank for reverse gear. From these two eccentrics, rods are led forward and fitted, the foregear one to the top, and the back gear one to the bottom of a curved link. Centrally placed within this link, and actually fixed to the end of the valve rod itself, is a die block. As the link can be moved up and down this die block by forward or backward movement of the reversing lever, it follows that three important positions can be obtained. When the link is in a central position the oscillating movements of both eccentric rods merely pivot and rock about the die block, without any motion being imparted to the valve rod. The engine is then said to be in neutral or mid-gear. If, however, the lever is pushed forward, lowering the link to the bottom position, the die block is subject to the full movement of the foregear eccentric. Conversely, the opposite result is achieved by fully lifting the link to bring the back gear into play. As a further point of refinement, any position selected between mid-gear and full gear, either in the forward or backward direction, gives a gradation from minimum to maximum travel of the valve. This control of travel is used to vary the amount of time the slide valve is actually open to admit steam to the cylinder. Technically it is known as varying the cut-off, and the driver uses it by giving full travel when hard work has to be done, but notching up and reducing the valve travel, whenever the going is easy. When the valve cuts off its admission to the cylinder at 30% instead of 60% of the stroke, the steam has to push the piston for a longer part of the stroke without any further help from the boiler, and is so used more expansively. In other words it gives up more of its pressure and escapes to the exhaust with less waste energy. So far as steam ploughing was concerned, drivers do not appear to have done a great deal of 'notching up' simply because it tended to make the engines a little sluggish. The flywheel, which is perhaps the most fascinating feature to any onlooker, is keyed to the left-hand side of the crankshaft, and its function is to keep the rate of each revolution as even as possible. It does this by storage of energy which carries each complete rotation over the two moments where the piston reaches the ends of its stroke and has to stop before reversing direction. As a matter of interest, the flywheels on most plough engines had elegant curved (dished) spokes in order to give clearance from the huge driving bevel gear wheel at the top of the upright shaft which drove the drum. The regulator handle is placed handy within the driver's reach, and is connected by a long rod to the main steam valve inside the cylinder steam space casting. The ploughing engine driver usually kept his hand on the regulator handle, although when manœuvring into a special position, it was best if he could set the engine running at the right pace before taking the steering wheel in both hands, especially on single engines with tank steerage. The tender attached behind the firebox provided, on top a stand for the driver, and below this a tank holding 150–225 gallons of water, and an open bunker above for about 5–7 cwt. of coal. A large flat horizontal steel plate was bolted on to the left-hand side of the tender as a standing place for the steersman who was employed whenever the engine ran on the highway. Whilst at work in the fields, of course, the driver did his own steering. For picking up water from wayside streams or ponds, a water lifter was fitted to the outside of the bunker. It comprised a steam valve which directed a jet of steam through a hollow ball-shaped cavity in the pipeline, joined to the suction hose lowered into the water below. The action of the steam created sufficient vacuum in the ball to lift water from 12 feet below the level upon which the engine stood, and discharge it into the tank. Various forms of change speed gear mechanisms were used incorporating a train of toothed gear wheels. By means of a sliding pinion on the crankshaft the engine could be put in or out of gear. Two road speeds were usually available. One important aspect of gear changing with plough engines was that the engine must be brought completely to a standstill before any attempt could be made to disengage or select a different gear. This in itself was little more than an hindrance or limitation imposed by the gear design, but what was a tricky business was if the engine happened to have been brought to a halt on an incline, even a very slight one, there was the immediate risk, once the gear was disengaged, that the engine might begin to run down the gradient. Unless the driver was smart enough to speed up his engine revolutions and to slam the gear in at the first indication the engine was moving of its own accord, a runaway was certain.

What about the brake? That is the first question likely to be asked by the reader, but the plain fact is that very few steam plough engines were actually fitted with any kind of brake as a separate piece of mechanism. For the purpose of retarding the engines when running downhill, or when they were required to stop in any definite position, reliance was placed on the holding-back effect secured by simply shutting off steam, opening the cylinder waste water cocks, and pulling the reversing lever into mid-gear position, or a little beyond. Air and steam in the cylinder were then trapped against the advancing piston which had to take impetus from the road wheels to compress the mixture. This caused a braking effect on the engine, quite sufficient in everyday conditions to give the driver full control down hill.

From what has been said already it will be appreciated how brake power could be obtained only when the road wheels were connected through the gearing to the piston.

As a precaution against mishaps, it was the custom of many drivers to carry a log of wood on the end of a chain hung over the bunker side in readiness to drop behind the wheel if necessary and sometimes, if the gradient were downhill, the steersman got off and placed the log in front of the wheel before the driver disconnected his gear from the crankshaft before changing speed. So placed, the log served as an effective scotch. But sometimes things went wrong and the engine moved off with anything it happened to be hauling. If there were a bank or a thick hedge by the roadside, into which the engine could be turned before any speed was gained, nothing worse than a nasty bump or two and a breakage here and there might result. Failing that, and if the engine was already in a mad plunge, the crew usually jumped off, although there was a grave risk that any implement or vehicle behind would swing out and run over them before they could get clear. The engine usually ended up on its side, as a 12- or 14-ton engine will do if left to run free. Sometimes these downhill runaways were caused by a broken valve rod or by water getting into the cylinder and consequently bursting out of the cylinder cover. Any upset in the cylinder block which destroyed the ability of the piston to compress air and steam within the cylinder, meant all braking power was lost. There is an illustration which shows a Fowler 'Single' plough engine which was ditched after water got into the cylinder when it had its head down while descending a steep hill in the West Midlands, and the cylinder front cover was in consequence knocked completely out.

XI

Direct Tractionists at Work

We have already seen how in the early and formative years several attempts at direct traction had ended in unfulfilled hopes. Sometimes the mobile steam engine had been attached to the front of the implement, or alternatively the engine and cultivating tool were combined in one piece of machinery. The excessive weight of the steam engine had, however, baffled all efforts to combine good cultivation with economical working. The story of the prohibitive weight of steam traction runs through the years, thwarting the best endeavours of these brave anti-cable advocates. Weighty iron parts, combined with just over a ton of water carried in the boiler and tank, were like Sinbad the Sailor's Old Man of the Sea irremovably seated on their shoulders.

When, in 1858, long-wearing steel wire ropes gave the handy-to-get-about double engine sets a further advantage, this style of working gained full favour and inventors associated with direct traction ideas received a decided rebuff. All the same, they never entirely lost heart. Frequently at shows, or on home farms, somebody chained an implement behind an ordinary traction engine whenever the ground happened to be dry and suitable. A typical occasion worth mentioning was the Highland Show of 1869 when a 6 h.p. Thompson Road Steamer, running on rubber tyres, with its upright pot boiler making it look rather like a small fire engine, showed off by drawing two Fowler-made ploughs across the demonstration field.

One novel and interesting development on the direct traction side was a combined engine and implement in the particular form known as a digger. The era of these diggers began with the steam forker patented in 1862 by a Mr J. Morris, Essington, Stafford. Whether or not this digger was actually built is uncertain, but the idea of attaching digging forks at the rear of an engine was developed in later design and construction. Morris' 1862 digger was to have a rear-mounted shaft carrying eccentries connected to the upright arms of the forks, to impart an almost circular motion to the digging prongs. Each of the three sets of prongs was to enter the ground in succession, so as to even out the digging load imposed upon the engine.

From Staffordshire, the digger scene passed next into East Anglia, a land of wider open arable spaces, where one or two imaginative and capable agricultural engineers did some remarkably original work.

Thomas Churchman Darby had a smithy and agricultural workshop in the small Essex village of Pleshy, near Chelmsford. Although his engineering capacity was certainly limited, his ideas were big. His first digger appeared in 1877, built in co-operation with Messrs W. & S. Eddington who had a larger engineering works in Chelmsford. They were able to make the engine and heavier parts quite easily from previous experience.

A cultivating outfit of their manufacture had been tested at the 1860 Trials held on Folly Farm, near Canterbury. The machinery included two wheeled frames, upon each of which was mounted a portable type engine for turning by belt, the underslung rope drums, as well as providing power to move each frame forward along the headlands. While mounted on the frames, the road wheels were removed from the portables. About the same time this firm had also undertaken some experimental work with a steam mole plough.

The actual erection, however, of this first Darby digger was undertaken at the smithy at Pleshy, where it must have created no small amount of local interest. Mr Darby's creation was a largish machine capable of working on a wide breadth of cultivation, in this instance just over 20 feet. In order to cover so wide an area in the field, his combined engine and implement travelled crabwise on the highway, with the digger mechanism lifted and partly folded on its left-hand side. But for action in the field, the two pairs of road wheels were turned through 90 degrees until their axles were in line with the boiler, and the whole great digger machine proceeded broadside trailing the digger forks stabbing madly at the ground. A rearward girder carried a cluster of eight disc wheels which, rudderwise, formed the steering unit. All told it was an ingenious piece of planning and construction, and even if a bit elephantine, was a worthy brain child of Mr Darby. Eddingtons made two further diggers for him, this time with a smokebox and chimney at each end of the boiler, which was fired from a centrally-placed firebox on the Fairlie pattern.

McLarens of Leeds, then newly established, also built a Darby digger powered by one of their single cylinder 8 h.p. engines. This digger, which was to be considered much simplified, was exhibited at the 1880 Carlisle Show where it was justifiably looked upon as novel and a great attraction. As before, a bevel drive from the crankshaft operated a two-speed transmission to the road wheels and a single speed to the digger prongs. This $15\frac{1}{2}$-ton digger dug a 20 feet 7 inch strip to a depth of 6 inches and tilled 1.19 acres an hour. It had what were described as 'creepers' or gripping blades on the rims of its four wheels to ensure that no slipping occurred. According to contemporary accounts, it left a 'garden style finish'. One feature of these diggers was that they continued digging while turning at the ends, but it took them about $2\frac{1}{2}$ minutes to accomplish a turning movement. It was claimed at the Carlisle Show that a digger required less horsepower for its work, because of the reactionary or pushing forward effort obtained from forcing the digging prongs backwards and upwards through the ground. Whatever help was so gained, however, was most likely offset by having to force the prongs downwards into the ground at the beginning of their cut, and of course, in the extra power necessary to shift such a ponderous machine over every inch of the fields. In all, McLarens made four of these diggers.

Our old friend Frederick Savage joined the digger party by constructing four Darby diggers in his works at King's Lynn. The 'Broadside' one he made in 1888 incorporated several of his own improvements, including six instead of three digging prong frames. A hydraulic jack, fed from the boiler feed pump, was used to relieve all road wheels from the weight of the digging gear when it was required to turn them into their alternative positions for travel on the highway.

Later on, Davey, Paxman & Co., of Colchester, made for Mr Darby a lighter and more compact type of digger. Altogether, about thirty Darby-sponsored diggers were built and by 1891 the original price of £1,200 had been reduced to £600. Some of these later diggers were sold to overseas customers. In conclusion, it should be said that in spite of many modifications and the reduced purchase price, the diggers never became popular. They were probably overweight on any but the hardest of soils, and most likely heavy on fuel and water, and of course they were cumbersome and awkward to move on the road. On a wide open Canadian prairie with an hour or so's run without the need to turn, and few occasions when it was necessary to waste an hour making the machine right for road travel, they might have shown up a little better.

Just prior to 1900 Mr Darby began development work on a separate digger implement which could be attached to any ordinary type traction engine. The prototype was made in new premises which he had opened at Wickford, known as Stileman's Works. This separate digger carriage was driven through a rearward slant shaft coupled to the engine crankshaft and imparted, through bevel gears, a rotating motion to rows of circular feet which cut up the soil as they gyrated egg-whisk fashion.

Another digger exponent was Thomas Cooper, a man who had great inventive talent combined with unusual mechanical skill. It was said of him that he could turn his hand to almost anything in the engineering line. His first digger, built in the late 1880's at Great Ryburgh, Norfolk, showed his approach to the problem was to have a more or less standard type of traction engine, with digging mechanisms built on at the rear of the bunker. A power-driven crosswise shaft, followed by a system of cranks and links, imparted a circular motion to the digging prongs. These prongs functioned on the principle of the leading row cutting downwards like wedges to open up the soil, while the second or rear row shattered and scattered the clods as they were turned up. The first of these diggers was named 'Pioneer' and was powered by an 8 n.h.p. compound engine. Just as McLarens had joined in with the Darby digger enterprise, so now Fowlers became interested in Cooper's machines, and five of his diggers were actually built at Leeds in the late 1890's, using Fowler 10 n.h.p. compound traction engines.

Experience gained with the first few diggers appears to have been sufficiently satisfactory and encouraging to warrant the formation, in 1894 of the Cooper Steam Digger Co. Ltd. In 1899 the undertaking moved into new works on the Wisbech Road at King's Lynn. What was probably the second digger made there was sold in 1900 to Mr Pryor of Weston Park, Hertfordshire. When the National Traction Engine Club, of which I happened at that time to be Secretary, ran a rally in 1958 at Weston Park, Mr John Pryor mentioned the steam diggers used formerly on the estate. He also very kindly put me in touch with a retired employee who had actually worked on the diggers. I do not think that I can do better than reproduce Mr C. W. Cadwell's letter:

Maiden Street,
Weston,
Hitchin, Herts.,
23. 6. 60.

Dear Sir,

Re your enquiry as to the Cooper Digger which was on Mr Pryor's estate, I am sending you details of what I can remember of the diggers used by his Grandfather.
No. 1 was named 'Victor' and was new about 1890.
It was a 16 h.p. engine. Two speeds 4–8 miles per hour.
The gear change was by lever.
Double forks—9″ spit and 9″ deep.
When No. 2 digger arrived, 'Victor' was converted into thrashing engine.
No. 2 Digger arrived new about 1900.
It was rated at 20 h.p. and had three speeds for digging and road, 4–12 m.p.h.
Single forks. Gear changed by removing keys and moving wheel on shaft.
Engine over the water tank.
Coal bunkers on side and soot plate was between cylinder and firebox.
Two steering wheels, one each side in middle over coal bunkers.
Turn in own length—brake on each rear wheel.
Width of work 11 feet. Work about 12 acres a day.
Could be used for thrashing or sawing by backing engine to drum or saw bench and using an attachment flywheel.
Last used for saw bench.
I began steering No. 1 digger when 16 years. Collected No. 2 from Hitchin Station and drove it until the Ploughing engines came about 1912.
I hope I have helped you. It is a long time since I drove the

digger, but I finished on the ploughing engines. I am 82 years old. If I can help you any more I will be pleased if you let me know.

Yours truely,
C. W. Cadwell.

One characteristic of Cooper diggers was that by placing the front wheels closely together under the smokebox, it was much easier to turn them within a short radius at the ends of the fields. One later development in design was to have a special engine with cylinders and motion at the rear, rather like Howard's 'Farmers' Friend' plough engine. On these engines the stubby unencumbered boilers were shown off in nice clean lines.

Mr Cooper was particularly interested in the production of a quick steaming and efficient boiler. As early as 1899 he had tried a water tube type on a digger which also had the peculiarity of being powered by a single-acting engine, i.e. the steam pushed the pistons from the closed or upper end only, and the return stroke was made by momentum as with most internal combustion engines. Furthermore, the valve gear of this engine had the refinement that it was cam-operated. For short periods overseas trade was quite good; four of the five Fowler-built Cooper diggers were actually shipped abroad. Other orders from Ceylon, South Africa and Egypt were met from the new works at King's Lynn. Further improvements in design were made from time to time, as for instance, the lightweight and cheaper three-wheeler turned out in 1903. This digger had a return tube type boiler with firebox, firehole and chimney at the leading end. By this arrangement, of course, it was possible to seat the driver at the front where he had a clear view ahead, and could steer a straight digging course more easily.

Digger construction continued at Lynn during the immediate post 1914 period, chiefly to meet orders secured from Egypt. About two dozen diggers with the engine motionwork once more positioned on the top of the boilers, were shipped to Egypt. Only a limited amount of interest seems to have been aroused in the home market for these Cooper diggers, perhaps because on our dampish land the heavy work was better accomplished by cable tackle.

Another promoter of diggers was Frank Proctor of Stevenage, Herts. His forker type digger had a simplified arrangement of long upright digging arms operated by a cranked shaft fixed transversely across the bunker behind a traction engine. Burrells of Thetford took up his idea, modified it a bit, and made a few diggers.

In spite of all the clever invention and enthusiasm of those who placed their faith in diggers, nothing worth while ever came of great efforts.

XII

The Steam Plough in North America

Although England was the country chiefly concerned with the introduction of the steam engine to the cultivation of land, our cousins in the USA were not slow to observe what was going on and by their own methods set about doing the same thing. They chose direct traction methods as opposed to our general acceptance of cable machinery.

One of the first Yankee ploughing engines we hear about was that built by Joseph Fawkes in 1864 for exhibition at the Illinois State Fair the same year. It had a pot boiler, and although at the Fair it surprised onlookers by hauling ploughs across a field of sunbaked soil almost as hard as concrete, on wettish soils its chief failing was to get stuck fast. In 1869 Philander Standish of California invented a traction type engine which pulled a separate cultivating machine, equipped with gyrating cutters.

An editorial in the *Pacific Rural Press* of 1876 pointed out there had been more progress with steam ploughing in Great Britain than in California and to some extent this was due to the fact that attempts by such British firms as Fowlers, Howards or Aveling and Porter, to introduce their engines and implements into the States had been impeded by the prohibitively high tariffs protecting American industry from imports of foreign machinery. We know a few sets were swung ashore from ships which had crossed the Atlantic, but neither in the USA nor in Canada, did our cable tackle ever gain any measure of popularity.

Agriculture in North America during the latter half of the nineteenth century was on an entirely different footing from that of the long-settled British Isles. The great rolling western prairies with their rough grass and bush scattered terrain east of the Rockies, and wide open spaces between the mountains and the Pacific in California, were coming for the first time under extensive cultivation.

95

The spaces were immense and there were no small irregular hedged fields to dictate where and how the day's work of a steam plough should be confined. Moreover, the virgin soil presented good grip and support for the wheels of the heavy engines, moving over it as they hauled their ploughs. Some idea of the vastness of the land may be gained from an account of one large California ranch where horse ploughing teams set out at dawn to move continuously ahead in the same furrow line until midday, when a stop was made for lunch at the halfway mark. After a meal, ploughing went forward again towards the horizon until the end of the piece was reached by evening. The night was spent there before starting the course home the next day. Although this is perhaps an instance of exceptionally long-distance ploughing, it does indicate the immensity of some ranch lands. Builders of American farm steam engines had these conditions in mind as they designed their tackle. Even more so than in Britain, the farmers persisted in their demand for an all-purpose engine which could plough and thresh according to season.

One of the chief personalities in the USA steam engine business was Jerome Increase Case, popularly known as the Threshing King of the USA He started from scratch, eventually building up a large engine building plant at Racine. This plant, between its first portable of 1869 and the closing down sale in 1928, made no less than 36,000 steam engines, principally farm types. When the Case Co. went out of business, their factory was bought by Massey-Harris. Case himself seems to have been a bit of a character. He was certainly a traction engine tycoon. They say he had a beard reaching from ear to ear, coal-black eyebrows, steel-blue eyes, and in old age his long hair turned snow white. It is said also that in the early days he dealt with all complaints by making a personal visit to the customer. On one such occasion, after inspecting a thresher, he decided his machine was no good, called for a can of kerosene and set fire to it on the spot.

He was also rather economy minded and never tired of picking up bent nails in his workshops to be straightened out for further use.

The first engine specifically intended for ploughing left the Case Steam Plow Engine factory at Racine in 1890. It weighed $9\frac{1}{2}$ tons, had an upright pot boiler, and was a three-wheeler for ease in turning quickly. It drew ploughs and it is said that while

9. Fowler 14 n.h.p. single-cylinder plough engine after downhill 'runaway' in the West Midlands

'Oxford' 12 n.h.p. compound steam plough engine No. 67 as 'old time' exhibit at the 1950 Oxford Royal Show

10. Fowler plough engine on beach at Gallipoli,

Fowler plough demonstrated in 1916 to Russian
Government for cutting trenches for military

engaged on demonstration out west in Montana in 1892, a pack of coyotes (small wolflike animals) were so mystified by the appearance of this strange machine that out of sheer curiosity, they stalked it at a comfortable distance. Like Aveling and Porter, the Case Company had a very distinctive trade mark or emblem. Case chose an eagle commemorating 'Old Abe' the famous old bird which an American regiment had carried about as a live mascot some years previously. A splendid brass eagle was fitted conspicuously and almost full length on the smokebox doors of all Case traction or plough engines.

Other United States firms such as Reeves, Gaar-Scott, American-Abell, Huber, Aultman, Minneapolis and Port Huron, as well as several Canadian builders, built various types of traction engines intended for direct cultivation. These makers all tried hard to produce a good all-round style of engine, but never quite accomplished their object with steam.

As will be seen from the picture, North American engines were quite distinct in type. They usually had round-spoked and rather spindly-looking road wheels, while water was often carried in round metal containers, stuck on at the side of the boiler or elsewhere. The small bore cylinders were frequently fixed on the side of the boiler, almost as an afterthought. On many single cylinder engines the drive was through a disc crank located at the end of the crankshaft, on the opposite side to the smallish flywheels.

These American engines were usually of the high speed type, and by fast rotation, were able to produce a high horsepower from cylinders of small diameter. On the whole, the American engines look good enough and they must have been interesting machines to drive or to watch at work.

The ploughs which these engines pulled across those enormous fields were of the American 'Gang' type. Instead of all the ploughshares and their mouldboards being fixed rigidly to the frame as in English practice, they were individually coupled and hinged to a frontal frame. Sometimes a gangway extended over the top of the frame so that the fireman riding on the engine could walk along the front of the plough. By manipulation of a lever attached to each ploughshare, he could either lift or lower it individually. As the name implies, there was a 'gang' of ploughs. Where formerly Red Indian smoke signals had conveyed their messages, between about 1890 and well into the 1920's the prairie skies were smudged or pillared with columns of smoke rising up from steam ploughs as they laboured across the land, changing its face.

Something with a real North American flavour were the important Annual Steam Plow Contests, held near Winnipeg in Canada. Every steam engine manufacturer in the States or Canada, sent engines to contend for the coveted best ploughing prizes, with all the seriousness of business competition. The engines were sent well beforehand by rail to Winnipeg, in order to allow time for careful tuning up, prior to contest day. Tampering by rivals was a not uncommon occurrence, so on the eve of the day the crews slept in their clothes beside their engines. As Winnipeg water was alkaline (contained excessive lime), it was also customary to bring in supplies of good water by railroad tank cars.

The site usually selected was a piece of virgin soil upon which each engine's lot was marked out by a line of flags. Sometimes the course was as much as a mile in length. Watched by large crowds, huge engines weighing up to 18 tons panted along their furrow lines at these Manitoba Steam Plow Trials, which by 1911, had become the foremost agricultural event in the world. It is significant that in 1910 a Case steam engine took first prize, but in 1911 an internal combustion engine tractor made by the same firm carried off the honours. That seems to have been the turning point, the oil engine had won. Only a few years previously these North American oil engined tractors had been so temperamental and difficult to start, that farmers employed boys to keep them ticking over all night rather than spend half the next day trying to get them going. The Winnipeg contests were discontinued during the 1914–18 War and never resumed, and so closed a most colourful chapter of the steam plough age in North America.

Besides the cut and thrust of commercial competition at these shows, a fair-like atmosphere of fun and jollity was added to the whole proceedings by games of skill for the steam plough drivers. There was, for instance, the Teeterboard or giant heavy-timbered see-saw upon which a driver's ability was tested to balance the engine, and to regain the balance after alternately tipping the ends of the teeterboard to the ground. Another stunt was to drive the engine up an inclined wooden ramp and to crack an egg laid upon it without breaking it. A further refinement was to come back down the ramp towards a watch hung on a stake with its lid open, and to close the lid without as much as denting it. All these stunts, of course, called for first-class skill in controlling the engines. Men who could win prizes at these games could drive a ploughing engine anywhere under any circumstances.

Henry Ford, the motor magnate, first became interested in mechanical transport when, as a boy of 12, he saw a steam road locomotive. He certainly never forgot the occasion because he claimed it gave him the inspiration to develop automobiles. At the time when he was experimenting with an oil engine tractor for farm work, he paid a visit to one of the Winnipeg Steam Plow Contests, perhaps just to see where the merits lay between steam and internal combustion power, and it is perhaps a pity he never had enough steam in his blood to have a go at an improved type of steam engine for ploughing work. He appears to have been convinced, early on and rightly so, that the future lay outside steam.

XIII

Steam Power Implements

Throughout the whole era of steam, the tip up, non-turning type of plough and the turning cultivator were the two main tools used in Britain. Other implements such as harrows, pond cleaning scoops, rollers or mole draining ploughs were, of course, employed to a limited extent. As the information so far given about the tillage tools has been brief, more will now be said about these implements generally. Though the greatest and liveliest interest was centred upon the engines, their sole purpose was to provide power for the implements which actually tilled the fields, so the tools must not be overlooked. The two-way balance plough was reasonably satisfactory except that being balanced, it tended to lift out of the ground at the rear or working end during hard pulling or jolting on shallow tillage, or on light land.

Sometimes, an extra man rode on the tail of the plough to help weigh it down. An anti-balance plough was then invented (by Fowlers in 1885) to overcome this fault. This implement ensured that when ploughing in either direction, the undercarriage and axle moved ahead of the centre or balanced position of the main frame and thus, by anti-balance, transferred more than half the weight of the plough behind the two-wheeled axle. The chief mechanical feature was a pair of toothed pinions fixed one on either side of the undercarriage, and free to run in inverted racks fixed to the main frame. When the plough was at rest, its weight caused the pinions to take up the top or balanced points in the racks. But since the ropes of the two engines were hitched at either end of the carriage, it followed that as soon as the rope was pulled to begin a bout, the carriage moved forward until the pinions locked up against the fore ends of the racks on each side. This movement, of course, caused more than half the weight to be at the rear of the plough axle. At the end of each run, the plough naturally ran back into the central or balanced position, making it easier for the crew to pull down the lifted end in readiness for the return crossing of the field. Ultimately, a type of plough was sold which could be adjusted to balance or anti-balance as required by the nature of the work in hand. On deep tillage work, the ploughs held themselves hard down into the work without assistance.

Steering a plough was a roughish job. The best the steersman might expect was a straw-filled sack as a sort of countrified cushion tied on to the otherwise hard wooden seat. The tension on the hauling rope tended to reach the plough in a series of surges or jerks, as the outstretched cable whipped under the strain from the engine ahead. As the plough lunged heavily onwards, the five or six ploughshares bit into and overturned the protesting soil, with a continual ripping sound. A yard before the end was reached, the ploughman steered the furrow wheel up on to the land, then quickly turned the wheels in the opposite direction, just before the plough stopped, so that as soon as its direction was reversed by the pull of the other engine, the implement moved over, forwards, up the field and on to the next furrow course. The ploughman then got down, walked to the uplifted end which would now be almost touching the side cover of the engine cylinder, and immediately the other engine slowly began to haul the plough back, the upended side of the plough was pulled down. Not only had the opposite side of the implement to be balanced out of the ground but at the same time it had to be turned into line with the new section of ground to be ploughed, i.e. out of the old furrow course into the next, using the help of the steering as left when the steersman dismounted. This 'Pulling Down' as they called it was heavy work, even if two men rode on the plough and cases of rupture were frequent. As the engine at the other side of the field increased the power of its pull and the shares began to bite into the new cut, the slack rope was picked up and dropped into the carrier; then the crew nipped alongside and clambered back into their wooden seated positions, and so the work went quickly on. Steam ploughing, however, was never up to the arrow-straight standard of horse work. Much depended upon the skill of engine drivers to maintain a steady even pull as this helped the plough-man to steer straight.

The English or seaming type of five or six furrow plough weighing 3 tons was widely used in Great Britain. A working speed of 3 to 4 miles per hour was customary for above that speed the quality of work was poor.

This plough laid unbroken furrows, all evenly raised on edge as the farmers like to see their fields exposed to the breaking down action of the winter frost. A fault perhaps of this style of plough was its failure to cover completely surface stubble, straw or rank weeds, and indeed a good deal of autumn steam ploughing was left looking a bit rough with quite a fair amount of rubbish unburied. Pressers or rollers were sometimes chained at the rear of the plough to consolidate the finish of the work, but a trailing implement was an awkward thing to turn at the end and a general encumberance. Disc ploughs, functioning by sharp edged diagonally-placed discs, were much better for covering up surface rubbish, but were used only to a small extent with steam power.

It might be true to say that the British built two-ended plough was never a realy satisfactory tool, so far as finish of work was concerned. French balance ploughs, which are still used for direct traction in France behind tractors, are of a much better design. In any event, this type of plough, carrying the extra weight of an idle side all the time, could scarcely claim full merit.

In all, Fowlers alone made 200 different types of ploughs, although this wide range included many special types for overseas farms. The soils of France, Germany, Central Africa, Italy, Algiers, South Africa, Cuba, Hawaiian Islands, Philippine Islands, and Peru each called for different kinds of ploughs.

Having said so much about ploughs, it must at once be made quite clear that, in Britain at least, they were never the favourite steam implement; in fact the term steam plough is really a misnomer. When I was a boy in Lincolnshire, the sets which came into the village of Barkston were never referred to as anything else but 'Cultivators'. Farmers and people about the village used to say So and So was to have the 'cultivators', meaning a steam plough set from the Sleaford contractors.

The three-wheeled turning cultivator was the really famous implement for steam work. Known at first as a 'Grubber', a name which clearly describes the action of this heavy-tined tool, the cultivator has always held first place in satisfying the farmers' needs. Its great value was the bursting up of the heavy clays in summer, to expose the roots of couch grass (twitch as it is often called), docks or thistles to the killing sun. The breaking up of the dry and baked clay into great clods, sometimes as much as a man could lift, was a job which at that time only steam had the power to do effectively. By 1910, six hundred Fowler type contractor-owned sets were almost continuously and exclusively engaged 'cultivating' during summer and after harvest on a 'Done and Crossed' way of working. This meant the field was first worked one way and then gone over a second time, crossing the previous direction of work. For the engines the hard pulling was, of course, during the first time over when opening up the solid soil. The 'crossed' work was easier and could be carried out a little faster. The second passage of the implement at right-angles to the first cultivation, gave the sharp ends of the tine points a second opportunity to catch and cut any deep roots of thistles, docks or similar weeds missed the first time over. Additionally, the clods were given a really thorough tippling over and more were left with their undersides exposed to the summer sun. This type of work also had the advantage that it let air into the soil with beneficial results to the next crop.

Cultivators were turned easily and automatically at the ends by the simple yet ingenious method of lifting the tines clear of the ground. The cables of both engines were hitched, each to separate eyes of the 'Y' shaped drawgear, on the front of the cultivator. During a bout, the pull of the engine brought one arm in line with the work, while the other reached outwards holding the slack rope clear of the implement as it followed. At the ends, as soon as the engine in the rear began to pull back slowly, the arm to which its cable was attached slewed round sufficiently for the extension of the drawgear behind the pivot pin to pull on a heavy chain. This movement engaged the toothed mechanisms on the axle and wheels of the cultivator. As the cultivator axle was cranked, the result was to lift the body of the implement, and so raise the tines clear of the ground while the turning movement was accomplished. As soon as the implement was face-about, and while the opposite engine was hauling slowly on its cable, the steersman pulled back a long upright lever in front of the large wooden box upon which he sat. This released the ratchet lock and down 'bang' went the whole heavy cultivator back on to the hard ground. The driver across the field then opened up steam fully, speeding the forward movement in a series of lengthening lurches. As soon as the sharp wedge-shaped points at the ends of the tines bit down deeply, the hard soil cracked and swelled upwards with huge clods climbing a foot or so along the tine shafts before peeling away on either side into a general melee of broken, tumbling earth. The whole operation was a witness of the great power of the huge engine ahead, straining with every ounce of its effort to wind in the steel cable coiled below its boiler.

Cultivating work was carried out at between 4 and 5 miles per hour although in special soil conditions, a speed of 6 miles an hour was possible. When compared with horse work, this speed was fast.

In order to avoid the heavy shocks to which the tine points were subject, when the implement was 'dropped' after each turn, later cultivators were provided with an automatic cushioning device. A cylinder filled with a gallon of paraffin was fixed so that when the implement began to fall its rate of descent was controlled by a piston moving inside the cylinder only as fast as the paraffin could be transferred, through a small hole, to the other end of the cylinder. This control provided an effective brake on the lowering movement of the tool.

Cultivators were steered by a large iron wheel connected by a chain to the front wheel which had a single central flange on it for the express purpose of turning the implement out of the straight line of the hauling cable. This was necessary if hedgerows were of irregular shape, or where there was a pond, manure heap, or stack in the field and the cultivator had to be forced on to a course not directly in line with the pull on the cable; the men called it getting it into a 'goring'. A flat tyred wheel would have slithered over the top of the land during any such attempted deviations. Help in turning to right or to left against the pull on the rope was also provided, by locating the drawgear pivot well behind the front wheel so that the nose of the implement had certain freedom to turn about the pivot point.

The largest type of cultivator used in Britain weighed nearly three tons; it had thirteen tines which cut a course 11 feet wide, and a two-engine set with this implement cultivating eight inches to 12 inches deep could get over 25 to 40 acres a day according to the nature of the ground. Cultivating was mainly a summer job, knocking about summer fallows on their year's rest from cropping. After harvest, the engines went on to cultivate wheat stubble land. A field might well be done twice in a season, the first time in June and then again about middle July. By the second tillage, any weeds not killed at the first visit of the engines, would be re-exposed when the cultivators returned a month later. Quite frequently, however, the farmer grew a crop of mixed clover and rye grass for harvesting in June and then, instead of waiting for a second crop in August, he would have the engines in straight-away to break up the land for what was known as 'bastard fallow'. This meant that it was not twelve months fallow but was a com-promise aimed at not losing a crop by a full season's fallow. In Lincolnshire, this practice was common, and provided much work for contractors' ploughing tackle during high summer. Of course, it often happened that some cultivation got left until late October when the autumn rains had set in, and then it could be a mess. The sticky clay combined with weed or stubble rubbish to form bunging masses in front of the tines, until the whole cultivator became like a huge earth scoop. When this occurred, the steers-man had to engage the lifting gear to enable the implement to ride over the blockage. I have seen such late work when the finished field was scattered with low mounds marking each place where a stoppage had occurred.

Experiments were made at various times to evolve a type of implement which would do everything that was wanted in moving all the ground at one 'go', without having to undertake the double or 'Done and Crossed' work necessary with ordinary cultivators. The quest was for a tool which would cut all the ground and sever weed roots like a plough, yet at the same time open and scatter the soil as effectively as the cultivator when working on fallow and stubble land and with the same relatively fast rate of progress. Fowlers took the lead by demonstrating at the 1894 Cambridge Royal Show, a special two-wheeled implement. It had eight individually slung and double-ended turn-wrest ploughs. As this implement turned at the ends of the field, it pivoted on one wheel, and the cross shaft carrying the eight ploughs was rotated half a turn to bring down the other set of ploughs, for the return run.

The pairs of ploughshares were, of course, opposite handed just like an ordinary turn-wrest horse plough. Eight furrows was quite a lot to pull at once and in spite of the breadth of work accom-plished in some cases, it seems unlikely that any engine of the time could have pulled this implement on heavy clay land except during very shallow cultivation. Messrs Bomfords undertook some tests on their land at Pitchill, where they found it was possible to cover forty acres a day. In fact they seemed quite enthusiastic about this new hermaphrodite tool, but in spite of their favourable report and the many thousands of pounds spent on experiments, the bud never blossomed and the idea became abortive. Once again, it was an instance of not being able to improve on a machine which had the unassailable quality of simplicity, as exemplified in the sturdy and straightforward cultivator.

The cultivator was liked best in some countries overseas, for instance Egypt, where there was little weed growth on the dry cotton lands.

An important item with steam work was, of course, the steel wire cables. We have already seen how much trouble was given by the early iron wire ropes before the introduction of steel ones in 1857 by John Fowler. Steel ropes are strong and hard-wearing principally because there is a high carbon content in the strands and this gives them a remarkably hard-wearing quality, as anyone who has attempted to use a file or hacksaw on them will know. It is beyond doubt a wonderful thing that such a rope should ever have been invented. When used with ploughing engines, steel ropes withstood the intensive abrasive action to which they were sub-jected on sandy or gritty soils; when the field was hump-backed and neither engine could be seen, there was a terrific grinding action on the rope which bore hard down on the crown of the land. Then there were the stresses of constant coiling and uncoiling of the rope on the engine drum. On an average, the length of rope on each engine was about 400 yards, but where large fields were likely to be worked continuously, outsize drums with 600 yards of rope were sometimes fitted. Usual rope diameters were $\frac{3}{8}$ inch, $\frac{5}{8}$ inch, $\frac{3}{4}$ inch or $\frac{7}{8}$ inch; the last mentioned was more commonly the size chosen for heavy double engine tackle.

One of the biggest improvements to steel ropes was invented in 1879 by Mr John Lang and introduced by Messrs Cradock, wire rope makers. This new rope had the notable advance that the individual wires had a long lay in the strands. From the illustration of a piece of Lang's lay rope it will be seen the lay (or spiral of each strand) is one where the small wires, and the strands into which they are twisted, have a long lazy twine. This simple idea, known as 'Langs Lay' marked a big step forward in lengthening the life of steel ropes; by ensuring that wear took place evenly across the flats of the wires.

Now and again, of course, a rope broke, calling for prompt repair by splicing; a job that the crews undertook in the field with all urgency, getting the tackle to work again with as little delay as possible. I have heard old steam ploughmen say that their foreman used to get very cross and swear if the rope were not spliced-up within 25 minutes at the most. Foremen with a keen sense of efficiency arranged for each engine to have a few feet of old rope lashed to a hind wheel spoke as a handy supply of short lengths of wire for splicing immediately a break occurred. Having this bit of rope ready on the spot saved time in not having to go back a half mile or so to the van to fetch a piece.

Fowlers made their own steel cables, while Wright's Ropes, of Birmingham and British Ropes of Doncaster, to mention only two, manufactured steam plough ropes in quantity. The tensile, or load bearing strength of the ropes, was in the region of 110–120 tons per square inch, so it will be seen how strong the rope was made for steam work. The violent tugging which occurred whenever the implement stuck on a landfast stone, or big tree root, placed an enormous strain on the cables.

XIV

Steam Cultivation Contractors

As we have already seen, by the 1870's steam work had resolved itself mainly into a contractors' business, principally because two-engine sets were too expensive an outlay for any but the largest farmers or landowners. The double-engine sets of tackle were sent out from the contractor's yards in the spring with a foreman, two engine drivers, a plough and cultivator steersman, and a teenage boy who acted as cook. There was also, besides the two engines, a plough, cultivator or harrows, a two-wheeled 250-gallon water cart and the double-boarded, bunk-fitted living van on iron wheels, in which the crew ate and slept. Each set was allocated to a particular district big enough to ensure that there was sufficient work to keep it engaged continuously without having to travel too far between jobs. The season usually finished around Guy Fawke's Day, when the sets turned for home, and left behind the now wet and soggy fields. As much overhaul work as possible was undertaken in the yards during the winter while the engines and implements were weatherbound.

Some of the earliest contractors' concerns did not survive the depression of the 1880's, but others managed to keep going until better times eventually returned to British agriculture. So numerous were these local contracting businesses, some large and some small, that only a limited account can be given here of the histories and activities of one or two of the major undertakings.

In 1868 Walter Eddison and Richard Nodding, both qualified engineers, opened a steam ploughing concern in the quiet little village of Cowley, near Oxford; a city at that time separated from Cowley by several miles of unspoiled green country lanes. Little is now known of the early days of this business, although it is pretty certain that the 'Works' could not have been much more than a glorified blacksmith's shop, lit by oil lamps in winter when illumination was necessary. Paper work was obviously kept to a minimum, because an old living van was all they had for an office. One can imagine some of the quaint looking ploughing engines they must have used in those times. As the long-chimneyed 'Puffing Billy' type engines clanked their way out of Cowley, the farmers' men, women and schoolboys, in each village through which they passed, must have gazed with no little amazement. According to the Rev. R. C. Stebbing, these Eddison & Nodding engines were Fowler 10 h.p. Singles with 6′ 8½″ chimneys and numbered 959/960 and 975/976. The crews, too, must have had an exciting time grappling with all the teething troubles of bedding down this new machinery to land work. Each season's work must have been like blazing a trail as they learned all the dodges the hard way, with snags and troubles at every turn. Yet all the time the face of that Oxfordshire countryside was being changed considerably by this intrusion of steam, leading as it did to more and more arable farming. Grass was fast giving way to grain.

G. M. Trevelyan, in his *English Social History* mentions how, along with other influences, the gradual introduction of machine ploughing brought more fields under the plough in Lord Palmerston's England. He tells of Matthew Arnold revisiting in the sixties the Oxford hillsides where he had roamed twenty years before, with his friend Arthur Clough, it was not 'Bungaloid' growth that the poet had to bemoan, but the more innocent spread of cultivation:

I know these slopes; who knows them if not I?—
But many a dingle on the loved hillside,
With thorns once studded, old, white blossomed trees,
Where thick the cowslips grew, and far descried
High towered the spikes of purple orchises,
Hath since our day put by
The coronals of that forgotten time;
DOWN EACH GREEN BANK HATH GONE THE PLOUGHBOY'S TEAM,
And only in the hidden brookside gleam
Primroses, orphans of the flowery prime.

In all probability, of course, the white smocked ploughboy with his brightly brassed head-nodding team of horses, was not entirely to blame for the replacement of white blossomed hawthorn with blue headed wheat. It was extremely likely that some of this was the work of sooty-faced and greasy-clothed steam ploughmen with Eddison and Nodding's new 'Black Horses' on iron wheels.

Business seems to have flourished with the Cowley firm, because in 1874, they spent £700 on some cottages and buildings standing on 2½ acres of land at Middle Cowley. This site was used for their new depot. At the same time, they began trading as the Oxfordshire Steam Ploughing Company. Richard Nodding eventually tired of his active interest in steam plough hire however, and in 1883, he sold his £350 share to his partner Walter Eddison. Another business change came in 1888, when Walter sold out at Cowley to his brother John Edwin Eddison, a consulting physician at Leeds. The Eddison family had close connections with John Fowler and Co. at Leeds, because Robert William Eddison (1835–1901) was a partner of the firm. When Walter sold out at Cowley, he moved over to join his brother Frank in the steam plough contracting concern he had established at Dorchester. It is from this business that we must now introduce a man who left them in 1887 to make a great name for himself at Cowley. This was John Allen who resigned his partnership at Dorchester to become manager for the absent Dr Eddison at Cowley. During his training at Fowler's steam plough works in Leeds, John Allen had gained a wide and thorough experience of steam ploughing both in Britain and abroad. He was a gentleman steamploughman, who certainly knew his job and was absorbed heart and soul with the idea of steam on the farm, as this history will eventually unfold.

John Allen found the Cowley business in rather a poor way; it was touch and go whether it could continue in the prevailing depressed state of agriculture. His first job was to put the mechanical side on a sound footing by bringing in from Leeds his old Upper Fitting Shop Foreman, William Anderson, as Works Manager at Cowley. Anderson was a first-class man for the supervision of repair work needed on the contracting tackle, when it came in for overhaul.

An illuminating history of the firm by G. T. Launchbury was published as a serial in *Allens Activities*, the works magazine of John Allen and Sons (Oxford) Ltd, in 1952. Amongst other things the author tells how at first the works did nothing but service the firm's ploughing engines and equipment. For this purpose, they employed a few highly skilled craftsmen who had, obviously, to be first-class men because there were so few machine tools then available. Later they accepted overhauls on farmers' threshing engines, showmen's road locomotives, and so on; in fact any work that would bolster up the low profit margin income obtained from steam cultivation contract work was welcome. Business increased to such an extent that a new workshop had to be opened in 1893 and, as may be expected, a 10 h.p. ploughing engine, suitably modified with a special crankshaft carrying two fly-wheels, was provided as a stationary power unit for the workshop machinery.

In winter, the works' staff was augmented by the steam plough crews. The home season depot work trained these summer travelling gangs in maintenance duties so that while they were away in the fields of distant hamlets or villages, they could carry out practically all necessary repairs on a day-to-day basis. Under these conditions, the summer working sets were more or less self contained for mechanical needs. There would be the odd visit or so to the local blacksmith by the village green, but apart from that they kept going themselves. Although none of the crew was an indentured tradesman, it was surprising what some of them could do. The foreman, or 'Outriders' as they were known to very old hands at Cowley, could undertake most repair jobs, including a good share of the more tricky work during major winter overhauls in the yard.

The Oxfordshire crews, like those of other contractors, were fairly liberally littered with unusual and colourful characters. The rough, knock-about, almost gypsy-like life had a special appeal to a certain half-nomadic type of man. As was the national custom, the gangs were paid a basic weekly wage, but earned an acreage bonus of a few pence each per acre for the work accomplished by their set. Only a proportion of this bonus, however, was paid weekly; the balance was retained by the firm partly to supplement basic pay when wet weather kept the engines from working, and partly as a Christmas bonus. When the bonus was paid out a few days before Christmas, the chaps often went 'whoopee' and celebrated in good old steamploughmen style. Tales of what happened on those 'Acreage payment days' are still related by oldtimers in the works at Cowley, who recall with amusement the boisterous scenes when poor men were rich for at least a few hours once a year.

About 1890, in order to remove further the dependence upon steam cultivation contracting, the firm began to take an interest in the hire of steam rollers. This new side of the business was an immediate success. As was only natural, a man with John Allen's ability and individuality now wanted to be his own boss, so in 1897, he paid the absent Dr Eddison £13,000 for the Cowley enterprise. When this sale took place the rolling stock comprised ten ploughing sets, eight threshing tackles, six steam rollers and perhaps a couple of traction engines used on odd haulage contracts. At that time, customers who hired the cultivating engines, included Henderson of Buscot Park, Latham of Dorchester and the Rev. George Moore, Rector of Cowley. More and more general work soon came in and by 1900 repairs to other people's steam engines had become the major activity with steam rolling second and steam ploughing third.

So far as steam ploughing engine repair work was concerned, a feature of the Oxford company was the amount and extent of major renovation and modification they undertook. They made their own boilers and in fact did everything required to complete a heavy repair. Their original engines were, of course, bought from Fowlers of Leeds, with whom the Eddisons had been so closely associated, yet early in John Allen's days, we find that modifications made to these engines at Cowley had changed their looks more than a little. The coffee pot covers on the cylinder blocks were removed and a new pair of fully exposed Ramsbottom safety valves was fitted. The latter change was a big improvement on the Fowler-fitted Salter safety valves, so easily tampered with by drivers, who often screwed down the springs to increase the boiler pressure above its certified limit, to get a bit more effort out of their engines. Another big advance was to rebuild engines with lengthened smokeboxes. This avoided the use of the ugly extension brackets to carry the front axle forward. This alteration also improved the steaming of the engines, and provided more space for ash to accummulate without piling up against the bottom tubes when the smokebox began to fill at the end of a long day's work. The longer smokeboxes, too, made the engines look better, adding a proud 'sticking out the chest' appearance to them. Later on, Fowlers, like most railway companies, turned to the extended smokebox as a standard feature.

Other cultivating contractors or farmer owners, whose repair facilities were limited, used to send their Fowler 'Single' type ploughing engines to Cowley when it became necessary to renew

11. U.S.A. steam plough engine for direct traction work. Built by Gaar-Scott 1909. (Double tandem compound type)

Borsig steam plough with vertical engine and rope drum at rear. It has wide wheel rims for moorland work in Germany

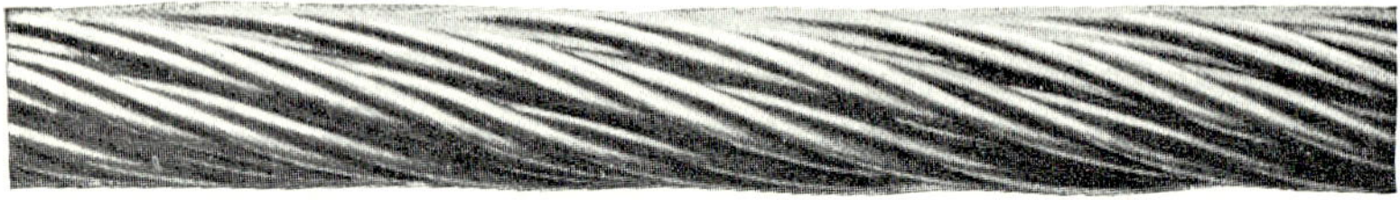

New rope

Worn rope

12. Lang's system of wire rope construction

[Reproduced by courtesy of John Fowler & Co. (Leeds) Ltd.

A fine pair of 1918 Fowler 'BB' compounds as seen just after
completion of boiler washout at River Witham Bridge, Barkston,
Lincolnshire in July 1939. These engines were bought by
Mr. H. J. Roads at Ward & Dale's sale in March 1939

the boilers. The 'Allen' boiler had fifty-five large 2″ diameter tubes, was pressed to 160 lb., and as it was made with $\frac{7}{16}$″ plate was altogether a sturdy job. Engines fitted with this type of boiler were well liked by drivers who said they steamed more freely in the field than the original Fowler ones. Testimonials came in the mail from satisfied customers; Mr J. Beeby of Rempstone, Leics., said the 2″ tubes made a big difference, while Mr Hailey of Stevenage, Herts, found that after his 14 h.p. Fowler singles returned home with 'Allen' boilers, each engine had burned 5 cwt. of coal less a day, and the set required two barrels less water. A different chimney was also fitted. This was the typical and unmistakable 'Allen' capped chimney made strongly in cast iron on a solid rectangular base, in place of the stovepipe type put on at Leeds. I think I prefer the stark simplicity of the stovepipe on a plougher, but it is a matter of taste. In addition to these larger and more significant alterations; new axles, wheels, crankshafts, link motions, slide valves, gearing etc., were fitted during the course of these extensive modifications.

Another feature of the Oxford Steam Ploughing Company was that they eventually combined contract and repair work with the actual construction of their own ploughing engines. In 1908 they turned out their first new 'Allen' ploughing engine. It was a compound, which with its comparatively high boiler pressure of 200 lb. developed 150 i.h.p. when at work in the fields. This engine was at once popular with the farmers since it burnt less coal and required fewer tanks of water carted to it each day. We must always bear in mind that as the farmers had the expense of buying the coal and lugging the water to the engines, they were keenly interested in hiring the most economical type. About 1911 a re-designed type of single-cylinder engine, having the same boiler pressure as the compound, but with a large diameter cylinder and piston valve steam distribution, was produced. Under test its performance was found to be almost equal to that of the 1908 compound. I have heard it said these improved single-cylinder engines were strong and hard workers. The last Cowley built compound No. 67 left the works in June 1913. Like a good many other ploughing engines, she was fitted with a drum drive engagement on her crankshaft, instead of this fitment being on the upright shaft.

After the 1914–18 war, about four new single-cylinder piston

H

valve engines were turned out. This marked the end of construction at Cowley. An improved piston valve compound plough engine had been designed early in the war years, but although some patterns and castings were actually obtained, no engine of this type was ever built. The steam engine had already come under the sobering threat of the motor tractor.

When the much dwindled steam plough hire finally petered out about 1938, the remaining five or six sets were either broken up or sold. Engine No. 67 was, however, preserved, 'just for old time's sake'. She still keeps company with a retired steam roller at one end of the motor scythe storehouse, where one late November evening, as dusk came on, I saw the sparrows come in, flitting and twittering to roost above her. There she stands, greased down, quiet and withdrawn in her corner, with the long years of noisy yet proud performances alongside the headland or hedgerows of far-away fields, now nothing but a faded memory of other and more glorious days.

The Eddison Steam Ploughing Co., Dorchester

Frank Eddison, older brother of Walter Eddison who was associated with the Cowley firm, started the Eddison concern in 1870, at the Dorchester Steam Ploughing Works. An interesting photograph of their yard in 1885 is reproduced. It shows ten of the twelve double-engine sets then owned by these steam contractors. From this picture may be gathered a good idea of what a contractor's yard looked like during the off-season. The engines, implements, living vans and water carts spent the winter in the open, and the long lines of upward pointing black chimneys must have been an obvious landmark. Covered accommodation was negligible, sufficient only to give protection from the weather to men engaged on very heavy repairs, when the tackle had to be stripped down, or doing jobs where it was an advantage to be near the blacksmith's hearth. All the rest of the work was carried out in the open, as may be seen from the plough engine in the right foreground with a hind wheel removed. Special attention is drawn to the exposed view of the friction drive inner wheel surrounding the axle on this particular engine. At that time it was common practice to fix the split and outer ring of these friction drives to the spokes of the road wheel, and when this was tightened down on to the inner wheel, the turning movement of the axle was transferred to the wheel.

An advantage claimed for these friction drives was that by slackening off the tension it was possible, when negotiating sharp corners, to permit the inside wheel to slip, and so compensate for the absence of a differential on ploughing engines. A dangerous shortcoming, however, was the tendency for the grip to slip when engines were descending steep hills. As we have already seen, the braking force was obtained by compression forces imposed upon the piston, and unless there was a firm grip in the train of driving gear between piston and road wheels, this braking power was lost. When such drive band slipping occurred, engines ran away, often with disastrous results. Eventually, of course, friction drive was replaced by removable locking pins in the hubs of the road wheels to give a more certain and positive drive, and to ensure full braking control.

Eddison's tackle toured the wide county of Dorset in the annual quest for hire, although there was naturally some territorial overlapping with the Oxford sets. By 1885, Eddisons were doing good business, employing no less than seventy men at their Dorchester yard and in the crews on the travelling steam plough sets. When John Allen became junior partner that year, a grand welcoming dinner (well loaded with speeches as was the custom in Queen Victoria's reign) was given for the firm's employees at the Antelope Hotel, Dorchester. In his speech, Mr Allen said, 'I have had good opportunity of seeing steam ploughmen all over England, and taking them as a body, I think that they are as good a lot of men as can be found anywhere'; this opinion he was well qualified to give.

Frank Eddison, who did not enjoy good health, died in 1888, aged only forty-seven. The business was then bought by a man named De Mattos, of Portuguese descent, who formed it into a limited liability company. Steam rolling soon became the major interest of the new company and the name of the firm was later changed to The Eddison Steam Rolling Company Ltd.

Ward and Dale Ltd, Sleaford, Lincolnshire

Most famous perhaps of all the cultivation contractors was the firm of Ward and Dale. A partnership of F. and W. Ward and W.

Dale was certainly functioning in 1890 at Sleaford, although it is very likely they were in business before that. Their premises were in Marholm Lane on the south side of Sleaford. This small red brick Lincolnshire country town stands right in the middle of an area of intensely cultivated farmland. Away to the east, the flat clays give way to the long expanse of the Lincolnshire Fenland with its thousands of acres of rich black soil, while to the westward lies the indulating country of clay, sand, ironstone and limestone lands, associated with the Lincolnshire Edge. The acid water of the fenland drains was bad for the boilers of cultivating engines, but there were plentiful supplies of good water in the streams and wells of the upland districts.

In 1908 the firm was incorporated as a private company, styled Messrs Ward and Dale Ltd, 'to carry on the business of agricultural steam cultivators' and taking over the old established firm just mentioned. When this changeover took place the company's capital was 250 ordinary shares of £100 each but only 210 were actually issued. The assets included the works premises and ploughing engines and implements valued at £15,000, the balance of £6,000 being working capital. They owned forty-eight steam plough engines, twenty-four drags (Lincolnshire name for a cultivator implement), twenty-four ploughs, twenty-four living vans and twenty-four water carts. This, of course, made up twenty-four working double-engine sets, and put the firm in the honoured position of being the largest steam cultivation contractors in Britain, and possibly in the world.

Ward and Dale liked to get their tackles out to work in March. To people living in the narrowish Marholm Lane, or in the little town over the level crossing, the departure of the first pairs of engines with all the clatter and rumble of their fourteen tons apiece, and huge iron-shod wheels grinding down the granite strewn surfaces of the roughish roads of pre-1914, was one of the annual signs that spring was at hand. These Sleaford sets went out all over Lincolnshire, as well as to parts of Nottingham, Northampton, Leicester and Rutland. The firm's name was almost a household word among the land working communities within a 30-mile radius of Sleaford. There was hardly a village thereabouts which did not have a pair of Ward and Dale ploughing engines do some work within its parish boundaries at least once a year. It was, too, an extremely wet 'back end' if a few of their sets did not remain out at work somewhere in Kesteven until early December.

Some idea of the amount of tillage accomplished by their tackles may be gathered from the following figures for 1914:

Harrowing	..	..	386 acres
Ploughing	..	..	9,521 ,,
Dragging	..	..	54,842 ,,
Total ..		..	64,749 ,,

With twenty-four sets in commission, the average amounted to no less than 2,698 acres apiece. An effort of this size was not achieved without many days of work beginning when it was scarcely daylight and continuing till after dusk. The high proportion of dragging again illustrates the popularity of the cultivator implement with the farmers.

I well remember how about that time, whenever a Ward and Dale set of Fowler 'Singles' thundered and clattered past our village school playground at Barkston, the foreman always walked about fifty yards in front of the first engine, pushing his bike. In accordance with the Locomotive Act 1865, the speed of traction engines on the highway was 4 m.p.h. on the open road and 2 m.p.h. when passing through villages or towns. The foreman jumped back on his bike as soon as the engines were out of the village street and rode on ahead of the clattering, jolting cavalcade always on the lookout for any horses which might be approaching. These animals could be relied upon to shy, if they had to pass the fearsome looking cultivator engines. If any appeared, the foreman signalled the driver of the first engine to stop and he helped, if necessary, to lead the horses past the stationary tackle.

The 'make-up' for travelling on the road was always the same. The first engine hauling plough and cultivator (in that order from the engine), was followed by the second engine with living van and water cart at the rear. Each engine was driven by its own driver, the foreman staking no claim to drive on the road, as I believe was the practice with some contractors in other counties. In parts of Northants, I am told, the foreman drove the left-hand engine (the one which pulled the plough rope on its left side) while on the road.

Ward and Dale's boilers were painted a warm horse chestnut brown, with the rest of the engine black, but smartly lined out in red and yellow. The implements were blue, living vans a reddish brown, and water carts green. The firm's name and No. of the set were painted in white at the rear of each engine's bunker. All their engines were Fowlers—it was a thoroughbred stable at Sleaford. The repair depot was well equipped and major over-hauls, including boiler changes, were undertaken as a matter of course. In addition they did a little general engineering repair work. At the end of each season they paid off the travelling crews with the exception of the foremen, who could if they wished, remain in the workshop.

As compounds came along, the stud was strengthened by these modern machines. The famous and well-tried old 'singles', how-ever, held on nobly for many a year and it was late in the day before they were all finally withdrawn.

During the writing of this history, I was put in touch with Mr Joe Rose of Frampton, who, as will be seen from the letter I received from him, was a former Ward & Dale steamploughman:

'Berleen',
London Road,
Frampton,
Boston, Lincs.,
24th March 1960.

Dear Sir,

I received your letter, thank you very much. I had four years with Ward & Dale. I started in 1915 untill 1918. I was first Cook and then Ploughman and then Driving.

In winter the plough crews picked up another job, some went back to the job they had left at Spring. But them what was inter-ested in their job always came back to the cultivating engines in the Spring.

Ward & Dale paid once a fortnight. The Foreman took in all Time Sheets and Acreage Book for work done on Saturday. Then on Monday when the men came back they were paid. The standing wage was paid up to date, but the acreage money was always a fortnight behind.

I hope you will be able to read my righting, I could answer a lot more if I was talking to you. Hoping this will help you.

Yours sincerely,
J. Rose.

I like that letter, not only because it gives us a contemporary link with the days of Ward & Dale, but also because it is obvious Joe looks back with interest to his younger days when he was out steam ploughing on the broad, black and flat acres of South Lincolnshire.

XV

The Traction Engine Challenges The Motor Tractor

1892 was the turning point in agricultural history when the Case Steam Engine Company of the USA made their first gasoline farm tractor which, for want of a better term, was called a gasoline traction engine. The idea of an oil driven tractor was also nourished in England, because in 1902, John Fowler & Co. Ltd. of all firms to do so, were experimenting with an internal combustion engine. Marshalls, the famous traction engine builders up on the Trent at Gainsborough in Lincolnshire, were also very early in the field with an oil-engined tractor, and by 1909 they had one on show which could pull a four-furrow plough quite nicely. By that year, too, one could say that internal combustion engined tractors were beginning to be seen up and down the country. Even if the infiltration was thin on a very wide front, steam had been challenged in what it had come to regard as its impregnable position as the one and only form of motive power for cultivation on the farm. How would it react?

Now the fundamental difference between the oil and the steam engine is the way in which pressure is obtained to force the pistons up and down the cylinders. With the oil engine, inflammable petrol or paraffin gases, or diesel gas oil fluid is led directly into the cylinder and ignited. The burning or combustion process builds up pressure within the cylinder itself, so this type of engine is known as an internal combustion engine. On the other hand, fuel in a steam engine is burnt externally from the cylinder, i.e. in the firebox, and the heat is radiated to the water in the surrounding boiler to generate steam pressure which is then led to the cylinder. The simplicity of the oil engine, without encumbrance of firebox or boiler, is at once apparent. So a threatening competitor had entered the lists.

By 1910 petrol-driven tractors were about in sufficient force to cause misgivings in certain circles well-wedded to steam. Our old friends the direct tractionists reacted in a positive way by assuming if these new-fangled 'tit-tit-hoppety-hoppety-tit-tit', smelly contraptions could pull a plough directly, the much improved light traction engines could beat them at that game. Nippy little steam tractors by Garretts of Leiston, Wallis and Steevens of Basingstoke, Clayton and Shuttleworth of Lincoln or Taskers of Andover, to mention only a few, were doing excellent work on haulage or timber dragging, so high hopes were entertained of the eventual outcome.

One of the more interesting attempts at direct ploughing was made by Wallis and Steevens who brought out one of their light steam tractors with its gears enclosed in an oil bath, and specially adapted to carry a plough body front and rear. There is an illustration of this engine which shows the general arrangement. The idea was that the engine should run backwards and forwards across the field, never turning, with the trailing end plough lowered, and the leading one raised as the combined engine and ploughs ran to and fro. The ploughman controlled the steering of the front wheels on the engine through his handwheel on either plough, and all the driver did was to drive the engine. With an arrangement of this kind, the driver would probably tire of running backwards with his engine half the time, and somehow too, it does not look right. All the same, it was a notable and imaginative experiment which proved that in eight hours, no less than four-acres could be ploughed at a cost reckoned at 3/3d. per acre.

In order to provide some reliable information on comparative performances, of steam and oil, the Royal Agricultural Society of England, obligingly came forward to sponsor some trials.

These were held from October 9–15, 1910 at Manor Farm, Bygrave, Baldock, Hertfordshire, in a district noted for its immense open fields over the gentle folds of chalkland, where the northern end of the Chilterns dips down to East Anglia. As it was autumn, the ploughing tests were carried out on an expanse of whitening stubble.

Conditions of entry were that the steam or oil engines should be able to haul an implement, drive a thresher, and pull a load on the land or on the road. In all, seven oil engines and three steam tractors entered as competitors. The Society provided the petrol

(Russoline at 8d. a gallon) and Welsh coal for the 'steamers'. Howard three-furrow type ploughs, set three inches deep and attached by six to eight feet chains to the tractions or tractors, were each manned by Howard's men. Several overseas visitors graced the occasion as spectators, while a notable Englishman present was Mr T. C. Darby of digger fame. Nobody was allowed to follow the implements lest over-eager onlookers should distract the crews. These important comparative performance tests were the first to be held officially in Great Britain between steam and internal combustion engines performing land cultivation tasks. On the final day, just over 400 people were in the field to watch the three steam traction engines, from various makers, complete their memorable direct haulage tests with their rather despised oil-engined rivals.

Manns' Patent Steam Cart and Wagon Co. Ltd., of Hunslet, Leeds, entered one of their 20 indicated h.p. compound light agricultural steam tractors; a baby of an engine weighing only $4\frac{3}{4}$ tons and priced at £400. Its principal features were the hind wheels placed closely together right under the large water tank which so conspicuously formed the rear of this engine, so that there should be no wheel tracks on the ploughed land. The boiler, pressed at 200 lb. was fired from the right-hand side close to where the comfortably-seated driver had a clear view forward. This unusual looking engine ploughed ·57 acre in an hour, at a cost of 4.03 shillings. Although no actual fuel figures were given, the consumption was probably on the high side, both for ploughing and haulage. In some measure this may have been due to the fact that the motion for the valves was inside the eccentric mechanisms, an arrangement which most likely did not give as good steam distribution as the ordinary Stephenson link motion. It must be said, however, that in this Mann engine we had an example of outstanding novelty of design. Her makers had made a brave attempt to produce a light steam tractor which would counter the threat of the oil engine.

McLarens of Leeds brought a £530 five-ton compound tractor. It had a boiler pressed to 200 lb., a smokebox superheater to dry and so increase the volume of the steam, as well as a feed water heater using exhaust steam to raise the water temperature to around 160 F. This was about the latest thing in steam tractor design, fast running, spring-mounted on all wheels and an extremely efficient engine. It was easily steered and driven by one man, and running at 5 m.p.h. ploughed one acre in an hour, at a cost of 3/11d. During the binder hauling part of the trials, a race was run between the McLaren and the Wallis and Steevens engine, both hauling two binders. So fast and furious was the pace, however, that the binders could neither cut nor tie the corn properly, and one spectator was heard to remark, 'What a ridiculous experiment'.

I suppose it was only natural the drivers felt like adding a little dash and fun to an otherwise well conducted ploughing and pulling match. The Wallis and Steevens came in first, proving that in addition to being the best looking steam tractor, she was also the fastest. The bulky flat-topped box housing the superheater on top of the McLaren engine rather spoilt her appearance.

The third steam contestant was, of course, the Wallis and Steevens Basingstoke-built engine just mentioned. She was a lightweight compound selling at £410. Unfortunately, however, her performance was slightly marred by persistent injector trouble. Yet even so, $\frac{5}{7}$ acre was ploughed in an hour at a cost of 3.66 shillings.

To those fortunate enough to be present at these memorable Baldock trials, it must have been an exciting and interesting spectacle to see the steam and motor tractors all lined up against the stakes, marking the two acre lots of the individual plots, and just waiting for the word 'Go'! One feature, which on such occasions always gives the steam engine its special appeal, is the upward curling smoke. Even when the engine is stationary, the thin plume of smoke as it rises, seems to give life and movement to the machine, adding also an air of mystery as to what is going on inside the iron horse.

It would not be quite right here to even attempt to dim the glory of steam by a detailed account of the motor ploughs. Let it be sufficient to say among those present were a Saunderson and Gilkins 45–60 b.h.p. 'Universal' and the smallish three-wheeled 20 h.p. Ivel paraffin type made at Biggleswade, similar to that now exhibited in the Science Museum at South Kensington.

The result of the trials was that the judges awarded first prize to the McLaren engine for best all-round performance. They had, however, no hesitation in saying the motor tractors, in spite of being beset by teething troubles, had shown up well. All in all, it had been a competition won by a well tried light steam traction engine just about at the top of perfection in this class of steam power, against a newish form of engine which, as the judges remarked, was 'still in its development stage'.

Quite apart from the Baldock proving trials, a good deal of other experimental work had been taken in hand by various manufacturers of traction engines. They were aiming at beating the motor tractor at its own game of hauling the plough; and in so doing hoped to produce a traction engine which could at last be regarded as a general pupose engine, in the fullest sense of the meaning on a farm. In 1911, Garretts of Leiston adapted one of their ordinary traction engines for direct ploughing work. Later on, in order to reduce the tendency to sink down into soft soil, they widened the rear wheels with outside extension rings, while detachable angle irons on the front wheels bit into the ground at the ends and assisted the steersman to turn the heavy engine. The spiked spudlets on the rear wheels were also lighter than the clumsy 'spuds' normally used on engines to give them a driving grip. Other aids were a belly tank to carry additional water and two-level drawbar. However, since this engine must have turned the scale at eight to ten tons, it most likely made heavy going in any but the driest of weather. In 1912, a Richard Garrett five-ton superheated light steam tractor was reported giving satisfaction to Mr T. J. Mugleton of Hammond's Farm, Stapleford Abbots, Romford. He said not only did it accomplish all his autumn ploughing with a three-furrow plough when the land was too hard for horses, but was also used to cut chaff, saw, grind, cut corn, fetch manure and do other haulage jobs.

In 1914 Marshalls of Gainsborough turned out a tandem type compound (i.e. both cylinders in line on one common piston rod) for use overseas.

Another maker of tandem type steam tractors, was John Maris Collings of Bacton Hall, Norfolk. He employed a mere handful of men in what was really a small agricultural implement business carried on in part of the outbuildings of the Hall at Bacton. We might say Collings was a freelance designer and builder, whose aim was to make light traction engines suitable for direct haulage jobs on farms. In all, he constructed only three such engines, using high pressure boilers (250 lb. sq. in.) which he bought from Alfred Dodman & Co. Ltd., King's Lynn. The only remaining Collings' engine has been preserved by Ronald H. Clark at Shotesham All Saints in Norfolk.

If we take a look at the general picture of the fortunes of direct traction, it has to be admitted that this way of doing the job never caught on in Britain, nor for that matter in Europe as a whole. The massive twin-engined cable sets held their own quite easily as the prime method for using steam power to plough or cultivate the land. Nevertheless, direct cultivation by steam traction engines, both large and small, did enjoy a fair amount of favour in many British Colonies. Over in the United States and Canada it was, of course, considered that the only acceptable course was to hitch the engine in front of the implement.

XVI

Steam Ploughing as a Pastime

My own introduction to steam ploughing came very early. For as far back as I can remember, I have a fairly clear recollection of the great ground-shaking two-engine sets of tackle going past our little red brick house at Barkston, Lincolnshire. This would be about 1912, when I was five years old. There was never any doubt when a set of engines was on the move in the neighbourhood because even if there was then little or no motor traffic to make other noises, the combined rattle, whirring and ringing of the meshed gear wheels and other general throbbing sounds associated with heavy ploughing engines, made them heard a mile away. As they passed, we boys used to run to the top of the yard and watch them, with the men riding on the engines bouncing up and down in a dither, in rhythm with the vibrations on the unsprung giants as they clanked on the way to their next job. Later on, after I started school, I was able to go out to the engines at work in the fields.

Nowadays, when organized cricket or football are taken for granted by every schoolboy, it may seem strange to look back not so very far when there was, as in our village, nothing of the kind whatsoever. Shortly after the outbreak of war in 1914, 'Billy' Archer, our schoolmaster at the little two-roomed school, joined the army, and since most of the young men were in khaki as well, there was nobody to organize sport either in school or out of it. The schoolmaster's wife took over the school, but planned pastimes were out for duration; even the Boy Scouts were disbanded while Major Bennett, the scoutmaster, was away with his regiment. We boys had to make our own diversions by begging cigarette cards from soldiers strolling out from Belton Park camp, gathering driftwood from the river Witham, looking for the nests of water hens, wild ducks or other birds, catching small birds in brick traps—or as several of us did in summer, spend some of our time during evenings or on Saturdays, out with the steam plough engines whenever a set happened to be working in the village. In those days, these massive, impressive and powerful steam engines were the thing, so far as machinery on the farm was concerned, and I think they appealed more or less to any boy with mechanical leanings. I know they arrested my interest and aroused an admiration which has been maintained ever since. So it was that during those war years and after, watching steam cultivation engines at work became quite an absorbing, happy summer and autumn pastime with me.

It was very seldom that any but Ward and Dale's engines came to work in the parish; although once I remember a set of green-painted Fowler singles with coffee pot tops on their cylinders which came from Morleys of Grantham to do some summer fallow cultivating at the Westfields Farm near the station. Usually it was the same Ward and Dale set each year—I forget the number of it now—but it was probably No. 7. The engines were Fowler singles and the set was in charge of an ageing foreman named Dick Stephenson. He always wore a peaked cap with thin black decorative braiding round the sides, an ordinary engineer's jacket, rather narrow stovepipe style trousers, and heavy hob-nailed boots whose only polish was the gloss on the dry leather obtained from constant rubbing on the big clods of clay as he went about his job in the fields with the engines. From the usual growth of stubble on his cheeks and chin, he was a man who shaved once a week, probably when at home during the weekends. In the mornings, he might be seen cycling around on visits to farmers to arrange future work.

He usually finished up about midday in either the 'Stag' or the 'Plough' to drink a glass of beer before returning to the tackle with a large, heavy, brown earthenware, one-gallon bottle of beer slung over his shoulders by a bit of rope, securely fastened round the neck of the old bottle. This roughish timetable permitted him to get back to the field in time to relieve the first driver, who then went for his dinner in the van.

Dick was a man of few words, because I never heard him say a lot, nor saw him put out or excited in any way, and so far as I can recall he did not swear much either. He seemed to accept us boys turning up uninvited at the engines, without any comment or warnings such as we got in other places from farm men, that unless we conformed to certain rules of behaviour, we should soon have the dogs on us and generally told to clear off pretty quickly. This tolerance at the 'Top' enabled us to spend what time we liked standing on the engines, where we had a grand-stand view of all that happened. It was always a thrilling experience.

More often than not the first farmer to have any cultivating done each year, was Joe Wadkin, who rented about 120 acres of arable land on the north side of the village. There was sand in the valley and heavy clay on the slopes of the Lincolnshire Edge. As soon as his first crop of clover ('Seeds' they called it) was carted off, he liked to put most of it down to bastard fallow, say, about middle June. Within a week of the last load going out at the gate, the engines would arrive, heralding their distant approach by the mellow ring of their cast steel gear wheels. This familiar and pleasant sound rolled gently and musically across the open Kesteven countryside. We could hear their coming quite plainly while in school, often long before they got into the parish. Sometimes, the first indication that the cultivators were ordered, was the sight of a red painted two-wheeled farm cart filled with coal standing in readiness just inside the field.

School finished at four, so as soon as we were let out, two or three of us would nip off after the engines. Old school pals I remember as my confederates in these expeditions were Albert Wheatley, Clarance Swain or Bill Millar. Actually it was better to go as a pair, because that allowed us an engine each to stand on in the field. I have a nasty feeling now that often we set off on these steam engine seeking expeditions without first going home for our teas and certain it is that on many occasions we did not return until supper time or later. Bicycles were not then possessed by boys at Barkston, so we always had to walk out to wherever the engines happened to be at work. The route taken by the heavy cultivating engines was plainly marked by their 24-inch wide wheels which ground down the loose granite surfaces, into a most distinctive trail. Of course, it was not really necessary to 'track' down the engines by following the wheel marks, because the light smudge of smoke drifting away from the field where they were working could be seen a mile away. There was, too, the far-reaching noise of the fast chimney exhausts, as well as the more pleasing and mellow bell-like tinkle 'ding-ding-dong-ding' from the teeth of the enga-

13. Fowler 'Superba' super-heated steam plough engine No. 17516 built for Mussolini in 1932 for the draining and reclaiming of the Pontine Marshes

[Reproduced by courtesy of John Fowler & Co. (Leeds) Ltd.

Mann lightweight agricultural steam tractor No. 1260 'The Green Man' built 1921 and used for direct haulage of ploughs. Owned by R. B. Haigh of Thaxted, Essex. On show at the Weston Park Rally, 1960

14. 5 n.h.p. Garrett light three-shaft single-cylinder traction engine
direct ploughing in Suffolk. (Note spuds on rim of engine wheel
to give a better grip and how steersman had to stand high
up in order to get a proper view ahead)

Pamplin Bros. of Cherry Hinton, Cambridgeshire, 14 n.h.p.
single-cylinder Fowler No. 3616 or 3886 after it had blown up
in Catley Park, Linton, Cambridgeshire, June 15, 1904
[*Photo:* Alan J. Martin collection

ging gear wheels, as each engine in turn moved a yard or so forward at the completion of a pull across.

I remember quite well a visit to some ploughing engines one hot June day in 1916. Whoever my companion may have been on that occasion, I do not now recollect, but we went straightaway after school to a set of Ward and Dale singles cultivating Joe Wadkin's Top Eight Acre. It was four fields up from the Lincoln road and the living van had been left at the top of the second field, a grass meadow known as 'The Sands'. We had to pass this van where the young Cookboy had his fire in a brazier outside, boiling a one-gallon kettle of water for the men's tea. It was better to have the fire outside the van on hot days, because those wooden vans soon got uncomfortably hot in midsummer.

The engines were cultivating on heavy clay, first time over after a crop of clover, and were already halfway up the slope. The drivers were really going all out, and after a few short puffs, as each engine took up its pull, it was worked full out.

The whole of the wooded slope above rang with the echoes of the violent barking of exhaust at the chimneys. As these singles were steered with a horizontal wheel on the left-hand side, a large iron plate was bolted on that side for the steersman to stand on when running on the highway. This made a jolly good place for boys to stand, and after hanging round the back of the bunker for ten minutes or so, just to test whether the driver was likely to be friendly or no, I unobtrusively climbed up on this perch. The driver was a roundish middle-aged man wearing a coned and broad brimmed straw hat (well speckled with oil spots) to shade off the heat of the sun. As I learned later, like all steam plough drivers, he stood with one foot raised up, resting on the side plating at the field side, holding the regulator in his right hand, and watching with intense concentration, the approaching drag (cultivator) as it bounded forward jerkily, with a little puff of dust blowing aside from it in the soft breeze. He was ready at an instant to shut off steam should there be any sign of mishap. Meanwhile, his sensitive hand was kept on the regulator, opening out slightly, or shutting down a bit, just as the speed of the pull waned or quickened. He and the engine seemed in unity, as if they had a common purpose in getting the job done; human hand was in harmony with steam turning the wheels.

While the engine was hauling in on its cable, the chief sensation

I remember was one of noisy intense effort. The single unbalanced crank lobbed round at about 160 r.p.m. and alongside it the two-valve gear eccentrics 'bobbed' eccentrically with that oddity of movement such mechanisms always display. The bevel gear wheel at the left side of the crankshaft meshed hard with the big horizontal bevel, whirring mightily as it conveyed the whole concentrated power of the engine to the rope drum below. On the outside of the crankshaft, left side, the dish-spoked flywheel, resplendent with its brightly-polished broad steel rim, spun fast in all its fascinating glory. Every now and again there came a surge in the pull, as if the drag tines had bitten down a little more deeply, or had run up against an extra hard patch of clay. It was then the engine hesitated slightly and rose up a little in the play on her hind axle and front carriage where the smokebox extension casting lay in its socket. She leaned sideways a little on the rope or field side, so that it seemed for a brief instant she might, under the great strain, use the implement as an anchor and slither out towards it; then, like a tug-of-war team which has wavered for a moment, she re-asserted herself in a new upward surge of power and on and on came the drag as she pounded away again. You could feel, through the combined frantic revolutions of the engine, and the heavy and jerky side-to-side movements, how hard the old Fowler was pulling at her cable. My main impression that day was the strength and determination of the engine to tear up the hard sunbaked clay of that rather out-of-the-way hillside field. Over and above all the other noises, of course, out of her shaking straight and plain black stovepipe chimney came the sharp fast chunter of the steam exhaust, which seemed to say, 'I-can-do-it, I-can-do-it', over and over again, as the smoke and odd tiny red sparks were thrown up straight and high into the evening sky. I felt quite sure this was engine driving all right.

As the drag came within about ten yards of the engine, the driver shut off steam a bit, gradually easing up until the engine stopped in a few expiring huffs with the drag rope shackle a few feet short of the rope drum intake pulley. Then the driver lifted the drum drive dog clutch out of mesh by using the long lever lying down the left side of the frame and immediately threw in the gear to the road wheels by using the lever for this purpose on the right side. The engine was then moved a few yards ahead into position for the return pull. This movement was made so quickly, I afterwards learned, to give the driver on the other engine, across the field, a visible indication that all was well for him to begin his pull. As soon as he saw our engine move up, he opened up steam and almost at once his rope became taut and there came a jerk at the shackle on the turning lever of the drag as the implement began to heel round for the return run. That over, the driver on the engine on which I stood, usually rubbed his hands on a piece of waste, as if by habit of relaxation, before either squirting a few drops of oil into this or that bearing, or throwing a few lumps of coal on the fire, or perhaps he turned on the water feed injector to replenish the boiler supply. This was the rest run, for while the drag was going back to the opposite engine, there was more or less peace and quiet after the shattering, gurring and grinding business of doing the pulling. Now the sounds were pleasant and wonderful in their softness and mellow tones. Down below the rope drum ran free, rumbling and tinkling merrily as the small pinion wheel was raced in a light-hearted manner, by the long ring of teeth round the top of the drum. Then too, there was the playful sound of the engine itself, with the regulator open just sufficiently to let a whiff of steam pass for the piston to slide backwards and forwards ever so smoothly and slowly. While the big end fell up and down in a casual manner on its crank pin, the soft exhaust of steam burbled and grunted kindly down in the bottom of the hollow chimney. It almost appeared as if the engine was saying 'I-can-work-when-I-want-to—but-now-I-am-at-play', and all this so softly and snugly and in such an endearing way while she was 'just ticking over'.

Another thing I recall from that occasion, was the engine passing under the branches of an ash ttee. This old tree stood in the hedgerow, where in the usual quiet of this field, it was much beloved by wood pigeons who would plummet down and with a final clatter of wings settle in its branches while they looked around before dropping to the ground to feed. Bit by bit, as the engine slowly gained headway up the field, we reached this tree and the chimney nudged into the thin drooping branches. Before the pull back was started, the smoke and heat had blackened and burnt a widish patch of the green leaves, but immediately the engine was opened out to pull in the drag, the roaring exhaust tossed the small branches and leaves hither and thither, blowing off odd leafy twigs and generally playing havoc with a fairly large patch of foliage.

How I came to go for a ride on the drag I cannot now remember, but I found myself perched up beside the steersman on his tool box seat as we lurched and bumped off towards the other engine on the north side of the field. It was rather frightening at first to see how violently the tines of the heavy implement tore up the ground, rooting great clods out right, left and centre. As a memorable experience the cake walk at the fair was nothing compared with the ride on that drag. The steersman could hang on to his large iron wheel for security, but I had to cling with my hands under the overlapping edge of the toolbox lid. This nearly ended in a mishap, because as we were turning and the other engine began to pull back it gave a sharp snatch at the rope; the steersman after leaning forward to pull the release lever, fell back wallop. How my fingertips missed being caught and nipped off when that lid went down with a smack, I do not know. Even now I look at them and think of that episode and how lucky I am they are there. At the time I said nothing about the matter, lest any further visits to cultivating engines should be forbidden.

Rising up towards the north west corner of our old village is a humped ridge of clay land, about 400 acres of it, known as Barkston Gorse. According to the old directories of Lincolnshire, it was not reclaimed from wild white thorn covered heath and woodland until about 1820. It is a great expanse of the toughest yellow clay, and as may be expected, never a year passed without one or two lots of steam cultivation was being attempted. A brook called The Beck, runs along the underside of the lower terrace of this clay land, and it provided a handy source of water for the engines, which, whenever they reached its banks, always sucked up a tankful. This saved the water cart horses the weary labour of lugging up the hillsides any more water than was necessary. There is a steepish rise from the Beck to the land on its north side, and for a short distance in one field, it went up so abruptly that at times the engines worked from the top of the field coming downwards bunker first. By so doing they ensured that water was kept covering the firebox tops, but on the really steep drop just above the brook, the drivers used their chained logs of wood, previously mentioned, as scotches to hold their engines from running back. It was amazing to me how the driver would ease his twelve-ton engine forward off the block, drop back slightly, then push the reversing lever forward, check it as it rolled down slowly, while he leaned over the bunker side dangling the block on its chain and finally dropping it just behind the rear right-hand wheel just where he wanted to position the engine for its next pull. Had anything gone wrong, the engine would have gone pell mell backwards down into the Beck, set deeply in grassy banks below.

Looking back, it seems a marvel that more accidents did not happen, because having to work those heavy old plough engines on shocking inclines without brakes called for the greatest skill on the part of the drivers to prevent accidents.

I well recall, too, how on those same hillside fields, the brightly polished brass bands round the engine boilers used to glint in the sunlight of a summer's day; we could see the sparkle of them from our house, a good mile away.

On one occasion in June 1917, a Ward and Dale set came to do quite a bit of cultivating right at the end of the parish in the far fields of Barkston Gorse. The engines came through during the afternoon while we were out at play, so as soon as the last school lesson was over, two or three of us went after them. To our surprise, the set was parked on the grass in a halfway field known as the Meadow Hills, through which the unfenced lane passed. In the far end of this field was a large pond, fed by a permanent stream, where both engines were backed up side by side in readiness for boiler washing out. The fire had already been thrown out of one engine, and soon after our arrival, the foreman told us to stand well out of the way. A driver then took a large longish-handled ring spanner, which he placed on the square of the scum or blowing-down cock, low on the right-hand side of the firebox. With a long piece of tube over the end of this spanner, and standing well back of the rear wheel, he opened the cock. A screeching, dense jet of steam and water spewed out immediately, like a glorified horizontal geyser and continued for ten minutes or so until the boiler was blown down and empty. As soon as the steam cleared away, a wide, fan-shaped area of grass could be seen completely covered with dirty white sludge and mud, as if white-washed by a giant. The engines had apparently been worked far too long without stopping work for a boiler washout, as frequently happened during good weather when more piecework pay could be earned by letting the boilers 'go'. A flexible hose was then connected to the injector delivery pipe of the other engine, which was still in steam, and the jet of water from it was used to swill out the water spaces of the first engine. Long pieces of wire rod were passed into the mud holes, to loosen and rake out the remaining sludge in the bottom of the boiler barrel, and in the water spaces around the firebox. The boiler had become completely sludged up for several inches above the foundation ring, easily understandable when it was known what dirty water was often supplied to engines on some farms from ponds or beds of streams with mud at the bottom. After finishing the one engine, the men packed up for the day and we had to go home without seeing the process reversed on the other. In the morning the men's next move would be to fill up the boiler of the empty engine. The usual method was to bye-pass water from the injector boiler feed on the other engine (which was in steam) and fill the washed-out plougher. As the water so injected was reasonably warm, it was not long before steam was raised again, especially if the fire of the engine in steam was shovelled into the other firebox. That was the only occasion on which I saw the boiler of a steam plough engine washed out, although it was quite common practice for passing sets to draw off the road on the disused sheep dip site near our Witham bridge on the station road, in order to wash out. Washing out a boiler is one of the most unpleasant and unwelcome jobs associated with steam engines.

The next day, the weather was fine and hot, so after school, Albert Wheatley and I set out to go and see the engines at work. It was a long dusty trail for nearly two miles to where they were cultivating clover stubble in the Far Speller, a twelve-acre field on the village boundaries, out of sight, on a west facing slope. The final approach was along the Green Lane, a wide grass track leading down to Hougham. This lane was pretty with sloe, thorn and crab bushes overgrowing the outer verges, and tumbles of honeysuckle or wild rose hung from both bush and hedgerow. The living-van had been left in the lane and the cookboy had his fire going in a brazier upon which was a huge two-handled cast iron cooking pot. He took the lid off the pot to show us the biggest lump of salt bacon that I ever saw boiling anywhere before or afterwards.

Of course, a gang of four healthy men and a boy working hard from sunrise to sunset on those wide open fields and hillsides of south Lincolnshire, ate like horses. Working with steam ploughing engines was a healthy occupation. The men worked always in the open air, breathing in the wonderful tang of newly-broken earth, and additionally, when cultivating clover stubble, there was the heavenly aroma of clover hay still clinging to the ground. The engines, too, had a peculiar yet pleasant smell about them. It was something a little undefinable, perhaps a mixture of the sweetness of boiled or baked oil, as the drips fell from bearings on to the hot sides of the boiler or firebox, combined with a peculiar scent which came from the odd whiffs of steam escaping from leaky joints. As the water used in the boilers usually came from ponds, ditches, brooks or rivers, it contained a good deal of minute animal life or plant organisms, and I can only suggest that it was this 'brew' in the boiler which gave that slight yet happy stew perfume to the steam. The objectionable fumes and smelliness associated with any form of oil engine-driven tractors, was entirely absent with steam. Many a man has remarked his pleasure of not only being able to see a steam plough engine at work but also his delight in the subtle and endearing smell of it.

Just before our arrival that June evening the engines had finished the field, which had been done twice over, as was always the case with cultivating work. The engine which had been down the slope was labouring its way over the uneven sea of clods up to the one gateway. Soon the first engine with drag attached nosed out into the Green Lane. The foreman helped each driver to hold the steering wheel, as his engine came over the rough, hard and humpy ground in the gateway.

With these old singles with their horizontal steerage, it was a matter of being extremely careful on rough going, otherwise the front wheels would fall aside into a hollow and spin the steering wheel almost as fast as the flywheel. Some men have been known to get cast into the air when that happened. Another really odd feature about this steering was that it was opposite-handed to the direction required. One turned the wheel to the right if one wanted to go left with the engine.

There had been some sort of disagreement between the two drivers while the field had been cultivated; one apparently held a poor opinion of the other's abilities as a driver. I rode on the latter's engine as the two moved off eastwards down the lane past the Four Cornered Speller on which a fine crop of wheat was growing, to the Three Cornered Speller, which was another clover stubble field next on the list. Now this triangular field was a big one of about fifteen acres; it had one short side and two very long ones, both running downhill. Having decided to begin work on the north side by the Speller Plantation, the two engines set off up the hill following the short hedge. The drag was behind the right-handed engine and when we got to the top, in the corner of the wood, the other engine was turned round and placed against the hedgerow headland in position for beginning work. The end of its rope was then pulled out and attached to the drag coupled behind the engine which was to proceed to the bottom of the field, uncoiling our rope as it went. The idea of taking the drag down to the bottom at the same time as the rope was pulled out, was to ensure that as soon as both engines were positioned on either side of the field, the top engine could pull back immediately on the first run with the implement; saving about five or six minutes in starting work. Away downhill went this engine, with the drum of the stationary engine sailing round in a carefree tinkle as the rope ran out.

The injector of the engine on which I rode up the field had been troublesome. It had been blowing off and needed to be reset at frequent intervals. As the other engine rolled off down the long slope, my driver got down to tinker with the injector, tapping it slightly with a hammer to release any bits of loose scale which may have been obstructing the clack. I had been standing beside the left-hand rear wheel, quite close to the running rope while he had been doing this. When he eventually climbed back on to the engine to try the injector again, for some reason or other, I moved back beside the bunker. Now while all this had been going on, the driver had failed to notice that the rope was almost completely unwound off his drum. Suddenly there came a shocking metallic snatch and thump, and before you could say 'Jack Robinson', the rope end had ripped open the shackle holding it to one of the four heavy cast steel spokes on the inner frame of the drum, snaked out of the guide pulleys and sprang viciously outwards and upwards before landing about twenty-five yards downhill. As the empty drum, impelled by the reaction of the breakaway snatch, trundled round freely in the reverse direction, I thanked my lucky stars I had moved back half a minute before that mishap.

The events just described took place during the First World War 1914–18 when agriculture was booming. The German submarine menace threatened British food supplies, corn prices were high enough to encourage every farmer to grow all he could; and in consequence, steam plough sets were much in demand. Farmers competed to get them on to their land first; the waiting list was usually long and it meant there was every encouragement for the crews to work all the hours the good Lord sent. Trade had never been so good. If the engines finished a job on Barkston Gorse at 10.30 p.m. at night; the set would be ringing its way up the West Street by 6 a.m. next morning on the way to another job.

It was about the middle of the War that the first motor tractors came into the village. Queer looking things they were, too, not altogether unlike traction engines with their big iron wheels. These tractors came on contract work at the direction of the County Food Production Committees with soldiers in khaki manning them; a driver for the tractor and a second man riding on the two- or three-furrow ploughs. These tractors often broke down, making them the object of ridicule by the steam plough crews working in adjoining fields. I have heard the drivers of the Fowler singles say, 'They'll never beat cultivator engines with them things. Just wait until they get on some really hard clay land and then you'll see what'll happen.' Steam in its pride still stood supreme.

During those times, when I spent many hours with ploughing engines, evidence of the war was plentiful everywhere, even in our remote country district. Headed by brass or bugle bands, long files of soldiers from nearby Belton Park marched wearily with heavy kits on training exercises. Overhead, biplane type aeroplanes were quite common as they flew slowly and low between the new Royal Naval Air Service or Royal Flying Corps aerodromes in Kesteven. Fields which had been 'worked' by ploughing engines were not the best of sites for a forced landing in a 'plane. During the autumn of 1918, a pair of Ward and Dale singles had dragged the Four Cornered Speller on Barkston Gorse, leaving it a tumbled mass of huge clay clods. Late one murky November Saturday afternoon, the pilot of an Avro 504K single rotary engine biplane developed engine trouble. No doubt as seen from the air, that field looked a big one and all right to come down in. Fortunately, the ash wood skids which were fitted to that type of aircraft in order to counter a tendency they had to 'nose' into the ground on land-ing, saved the craft from pitching headlong. But so rough was the going, it ran only twenty-five yards or so before coming to a stop. When the R.A.F. rescue gang arrived that night after dark, they were so discouraged by the cultivating engine clods, that they decided there and then to dismantle the aircraft rather than even try to wheel it to the gateway.

One autumn, just after the war, a set of Ward and Dale singles came to cultivate twenty or thirty acres of wheat and barley stubble land on Barker's farm, just up the hill above our house. The work there was on ironstone land, stained reddish brown by the iron-bearing marlstone five or ten feet below. It was early October and I well recall how several of us were with this set of tackle one evening until it got dark. Work went on as the hunter's moon loomed large above the dark outline of Minnett's Wood on the eastern skyline, and still the drivers pounded on with red hot sparks falling brightly into the ashpans of the shaking engines as they panted and strained at their cables. As we went home across the grassfield below, we could just see the moonlight reflecting on the white tails of the rabbits as they bolted down their holes in a warren they had made in a sandy patch of soil. Behind us the plough engines still pounded on with the work.

It has been said earlier that steam engines were prone to get bogged down. I saw only one bad instance with a ploughing engine. It was after the war, perhaps in the early 'twenties, when a set of Sleaford singles came out very early during some dry weather in March to do some sand land cultivation for Joe Wadkin. A twenty-acre field known as the Walks had been done over once and the engines were repositioning themselves for the second time over. The engine moving from the top near the Walk's Hovels, instead of following the hedgerow, cut a course down the field some thirty yards from the hedge. All went well over the apparently desert dry surface soil, until the engine was about three parts of the way to the Honington Road, when her hind wheels settled down as she dug herself to a standstill. Sleepers and other pieces of wood were brought in a cart to pack under the wheels, but no, they were pushed down or outwards and the engine was still unable to lift herself. Quite a number of people, some of them women, went to have a look at this old Fowler with her rear end down at a steep angle and she certainly did look in a pickle. The next attempt was made by digging out the soft wet sand and placing a long sloping ramp of sleepers and heavy timber ahead. The sister engine was then brought into position at right angles alongside the highway hedgerow and her cable led out to the one that was stuck fast.

When all was ready, the foreman stood where he could be plainly seen by both drivers and dropped his hand as a signal to 'go for it'. The two engines responded to their full regulators and slowly the 'stuck' one began to pound up the wooden ramp, like a heavy old elephant rising from sleep. All looked well as she mounted the top magnificently—only to fall down again as soon as she touched the naked ground. Straightaway she dug herself in as badly as before, in spite of the great help she was getting from the full pull of the other engine's rope. This caused one or two dirty hands holding uplifted caps and much scratching of heads. Then it dawned on somebody, one of the farm chaps I believe, that the engine was on a course directly over a piped land drain. If this drain had become blocked or was unable to carry away all the spring water, that would account for the 'quicksand' conditions of underlying sloppy sand the men had dug up, when making holes to let the timbers in under the wheels. All this happened during Lent, because I and other boys in the church choir, had to rush off to choir practice; even a stuck steam plough engine would not have been accepted as any excuse whatever by the Rev. Edward Mansfield Clements, our keen rector. However, after choir practice, we rushed back up the road, just in time to see another pull made. The foreman was helping the driver of the bogged engine with the steering and as soon as both engines went for it again, the one in the mud was steered sharply to the right, to take her off the drain course.

The dodge worked and up she came on to level ground again. It had taken two days to get the engine out and I heard the foreman say the strain on the engine pulling with its rope had broken a bearing casting into the bargain.

During my early experiences with those old Fowler 'Singles', I soon noticed an oddity in their speed gear arrangements. They had two speeds, a high one for the road and a low for field work, but only one of the two gear wheels could be accommodated in the driving position on the end of the crankshaft. This meant that before going into a field, the driver had to take off the 'high' wheel and replace it with the 'low' one. Whichever wheel happened to be out of use was carried on a dummy stub axle fixed on the right-hand side wooden footwalk. A similar arrangement can be seen on plate 9 where the 'high' gear wheel of the 'Oxford' engine rests on the footboard just below the high pressure cylinder.

XVII

The Steam Plough Owners Organize For War

One feature of British steam plough contracting activity had been its general lack of co-ordination or organization by the owners to protect and promote their common business. Competition was always keen. In consequence there was often a good deal of under-cutting in prices, which varied from county to county, according to the best bargain the individual owner could strike with his farmer customers.

After the outbreak of war in August 1914, however, the idea of forming some sort of association was discussed again in earnest. The plough engine owners knew that if the Kaiser's armies were to be beaten, more food would have to be grown in this country, and they naturally wanted the extra cultivation work for their beloved steam tackles, rather than see it go to the new-fangled motor ploughs. A champion and leader was found in Mr John Allen, of John Allen and Sons, Oxford, who was not only patriotic in outlook, but was also one of the firmest believers in the steam plough. A meeting was arranged on February 4, 1915, in the Cannon Street Hotel, when with Mr Allen as Chairman, it was decided to form the Steam Cultivation Development Association. Of the forty-six rules of the constitution, part of rule 2 read as follows,

The Association shall encourage, promote and protect the business of steam cultivation to the United Kingdom of Great Britain and Northern Ireland and generally obtain for its members the advantage of mutual co-operation and support.

The membership fee was fixed at £1 per two engine set. This inaugural London meeting seems to have given these country-residing contractors an opportunity also to let off some steam about the 'hard-up-ness' of their industry. It was said the charges were the same as they had been in the 1890's; that farmers under-stated the acreages of their fields, or often kept the tackles standing waiting for coal and water, and it often took twelve months before many farmers paid their bills. Another unexpected moan for 1915 was about the growing practice of farmers' horsemen refusing to work late on Saturdays or come out early on Monday mornings with the water cart. Some discontent about wages was also going around among John Allen's steam plough crews, who were threatening to strike the next month.

During 1916 the Association's discussions were frequently con-cerned with the manpower position, which continued to get worse as men from the ploughing gangs either enlisted or were con-scripted to swell the British citizen armies. The local tribunals had a nasty habit of calling up men indiscriminately, so much so in fact that there was a scarcity of skilled drivers and implement steersmen. One idea put forward in committee was that German prisoners of war, many of whom had steam engineering experience, might be drafted to work with the cable tackles; however, nothing came of this proposal. During the course of some committee talk about motorists agitating to get all steam traction engines off the roads, Mr Allen came down on them with the scornful remark, 'these people who have never so much as owned a horse or a don-key'. The County Councils too, were complaining loudly about damage to their road surfaces by the passage of heavy steam engines and were also wanting smooth wheels on traction en-gines, a £20 annual licence, plus a daily payment of 10/– whenever an engine was outside the county of its registration. One Council had gone so far as to suggest some sort of communication cord or chain should be provided to give signals from the rear implement or vehicle to the engine drivers, when ploughing sets were moving on the road. These were very stiff proposals.

At one meeting, Mr Pamplin, the well-known Cambridgeshire contractor from Cherry Hinton, said that in his opinion, it paid to have a fifth man with every steam plough set; otherwise up to 25% lost working time would result. Moreover, there had been one or two instances where the police had issued summonses for neglect of the rule that a man walk in front of tackles moving on the highway in order to warn people with horses. It was, there-fore, all the more advisable to have an extra man even if only to perform this duty. A tally made in 1916 showed there were about 450 twin-engined working sets in England, manned by crews whose average weekly wage was 15/-, plus 2¾d. an acre each time over. Some owners paid overtime rates, others did not.

As 1917 came in, the real seriousness of the war situation became more apparent. The rate of sinking of our food and munition ships by German submarines had gone up alarmingly, making increased production at home of the first importance. More acres had to come under the plough, and quickly. As Adviser to the Food Production Department, Mr John Allen had become an extremely busy man, often spending four to five days a week in London, besides managing his own business at Cowley. He was doing all he possibly could to persuade the Government that increased use of the steam plough was the answer to the agricultural cultivation problem. It is said that while in earnest conversation on the matter with Lloyd George, the Prime Minister, he became so enthusiastic that he told him a set of steam tackle could if necessary, plough up Piccadilly. 'Nonsense!' replied the Premier, 'I thought you were a sensible man.' However, in the end the Government gave way, by agreeing to place an order with John Fowler and Co., Leeds, for sixty-five double-engine ploughing sets complete. This victory pleased the old steam plough masters. A meeting was called on June 13, 1917, when Mr Alfred Fowler was asked to come along specially to discuss which types should be built under the welcome government order. He said 16 or 18 n.h.p. engines could be provided for the same price, but his suggestion that the 16 n.h.p. type was best for England was adopted. Following this an invitation was given for a small party of the Association's members to visit Fowler's works to decide a few details of design and see the first engines and implements under construction. This event took place in July, when after inspecting both A/A and B/B type engines being assembled, a suggestion from the visitors that 600-yard ropes ought to be fitted to all the government-sponsored engines was accepted by Fowlers. The first sets were promised in November and it was hoped to continue delivery at the rate of ten to twelve sets a month until completion in the summer of 1918.

In addition to the work he did inside the office of the Food Production Department, Mr Allen travelled around the country as much as he could, inspecting and advising on steam plough work wherever he went. His life and soul were in steam ploughing. While speaking to the Association members in the summer of 1917, he said, 'I have been a steamploughman all my business life. It has always been a pleasure to me to go into a field and see a pair of engines working a cultivator properly manned by men who know their job.' Those who know steam ploughing will appreciate the feeling behind that heartfelt remark.

Mr Arthur Stratton, of Alton Priors, Wilts., crystallized the outlook of the Association by saying, 'The steam plough owners are the navy of agriculture and Mr Allen the Jellicoe of the steam plough'; such were the sentiments of the time. In the summer of 1917, some steamploughmen were released from the army to return to the fields with the engines. One Association member soon complained that he knew of an instance where six men had returned from France, only to be promptly called back to the colours by the Local Recruiting Officer. As usual, it seems bumbledom had to have its little bungle. The urgency of the war effort had, however, softened the attitude of the police, who no longer objected to living-vans by the roadside, nor did they attempt to enforce the 1865 Locomotive Act requirement that whenever an engine stood working within twenty-five yards of the highway, a man must be stationed on the road to signal the driver to stop immediately any approaching horse showed signs of becoming restive.

Trouble with the government-sponsored sets did not end with the placing of the order, because they now began to haggle about the price to be charged to purchasers. The Government were asking £3,500 a set, which seemed too high a price even for the contractors to pay. After much coming and going, Mr Allen was eventually able to make a bargain at £3,000 a set, payable on the instalment system. The Food Production Department on their side insisted that the minimum amount of land cultivation by all the steam plough sets should be 2,000,000 acres. If private contractors could not afford to buy up all the sixty-five sets they had ordered, then the Local War Agricultural Committees would take them over and operate them on hire contracts to the farmers, along with the motor ploughs already under their control.

There was a shortage of coal at the beginning of the 1918 season, so it was with some relief that the members were eventually informed that 50,000 tons had been ear-marked for steam plough work in Great Britain. Scarcity of food, which was already rationed, was another problem. With his usual forcefulness of opinion, Mr Allen promptly told the Ministry of Food that steam-ploughmen customarily ate twice as much as ordinary men, and no doubt he was able to obtain some concessions in the way of odd extras for the gangs. Labour continued to be a problem; never before in the history of the industry had men moved about from owner to owner as now. In many districts, this drifting about of men had previously been discouraged, by owners refusing to take on a driver or ploughman who was trying to change his boss. But now men wished to move more freely in order to bargain for a bit of additional pay. It was noted, too, that in spite of the weekly wage for steamploughmen in Essex having risen to 36/- plus 10d. an acre bonus; men were leaving their life on wheels with the tackles for highly-paid bench jobs in munition factories. At an Association committee meeting in March, 1918, it was suggested that perhaps a steam ploughmen's training school ought to be set up, but this did not find general favour with many of these old-fashioned masters. They thought, as always, that the only training ground was working with the tackles, learning the job the hard way so to speak. Mr Percy Grundy, the notable steam cultivation contractor from the Glendon Engine Works, Kettering, began to accept the new conditions which employees had come to expect, by offering a bonus to each of his men who found a new man for his tackles. Incidentally, Percy Grundy had risen from chief clerk to the late Alderman Wicksteed, to take over from him in the early 1900's, the steam plough side of the business. The name of Grundy was sure to be mentioned in any steam ploughing talk over a wide area around Kettering. To his former boss, Alderman Wicksteed, stands the credit for the invention of the boiler tube expander, a tool subsequently used universally both for traction engine and railway locomotive boilers throughout the world. We are all indebted for this bit of information about tube expanders to Mr E. E. Kimbell of Boughton, Northampton, who with his early days of steam plough driving experience now well behind him, is always pleased to keep green the memory of those impressive years of his life. There is nothing he enjoys more than a glass of good beer and some steam plough talk to go with it.

The year 1918 was, perhaps, the last good one for steam work—it was the Everest of the steam plough era. All the owners were busy, and a pleasant air of prosperity pervaded the whole of their

15. 'Old time' demonstration of steam ploughing. Grantham and
Colsterworth Farmers Ploughing Match, October 5, 1957, at
Welby Pastures Farm. Ex-Ward and Dale Fowler engine
No. 15154 now owned by Henry Warner of Welby

[Photo: Grantham Journal

Fowler 'AA' No. 14726 owned by Mr. H. Jackson at work
cultivating in July 1960. (Note especially the strain on the taut
rope, the dents in the cylinder casing due to pulling the plough
too close to the engine and the typical 'steam ploughman's
stance' of the driver)

[Photo: Barry J. Finch

16. Lightweight cable steam plough built in 1927 by Bomford Bros. of Pitchill. Sentinel type engine and boiler. (Note vertical winding drum)

Crankshaft and gear wheel pinion. John Patten's Fowler 'AA' No. 15362 'Lion', June 1960

cultivating activities. The smoke of their restless engines went up in a thousand columns over the fields of this country. I met a steam plough driver during the middle 'thirties, working with a set of compounds at Elton, alongside the Peterborough–Oundle road, who while commenting on the general running down of steam work, recalled the war days when on one occasion in 1918 he had counted the smoke of no less than twelve other sets within sight of his own. That in itself speaks for the amount of steam work done in the last victorious year of the war. Although most cultivation contractors now had good bank balances, one man, Mr John Allen had considerably overdrawn on the balance of his energy and health. His fervour for the steam plough during the war had so spurred him on in the national effort, that he had seriously undermined his health.

When the 1919 season began, he was rather concerned about the possibility of an eight-hour day, with the serious effect that this might have on steam cultivation contracting which, outside the winter months, had always regarded its normal day as every hour the good Lord sent them between light and dark. After the winning of the war in November 1918, there came, too, a change of heart by the Government, which somewhat naturally lost interest in steam plough engines. This change of attitude must have seemed a bitter anti-climax to a great steam-minded man like John Allen. It was also a great disappointment for him when the motor plough he disliked so much, and which he had so nobly resisted all the difficult war years, began to gain favour over steam. On his doctor's advice he took a rest in his native Northern Ireland to recuperate. Unfortunately, he never fully recovered from the strain of his work for steam ploughing during the war years. Nevertheless, he came back as chairman for a few years, continuing to ridicule the chances of the tractor; and encouraging the members with his high hopes for the future when the temporary agricultural depression had passed. At one association gathering in 1921, he mentioned that several small local meetings had been held, 'with lunch at half a dozen pleasant country towns'. One can imagine those bowler-hatted contractor potentates, with brightly polished boots travelling by train to the venue, looking out of the windows, as the country-side rolled by, to watch for this or that steam set with its pair of engines working along hedgerows in pretty rural settings. If the engines had moved on, their 'trademark' in the form of 'last time over' wide wheelings wandering along the hedge sides, would plainly mark land that had been cultivated by steam. What grand talk there must have been, too, at these little luncheon parties in old hotels! As those seasoned old hands of steam plough owners' rearguard looked across the table, just think of the live steam tales they told, how the work was going on with their own sets, what the foreman or cookboy had said, and so on—all of it first-class current steam cultivation conversation, never to be heard again.

Our Mr John Allen, J.P., O.B.E., M.I.T., carried on bravely as chairman after he returned from his 1919 convalescence. At the A.G.M. held on April 23, 1925, in the Holborn Restaurant, London, he looked down from the chair over that gathering of old friends and colleagues, in the steam plough game, and in a mood of reminiscence, told them, 'for some reason or other, of all forms of machinery which appealed to me as a lad, the steam plough was the one I took to in the most kindly way'. He was thinking, of course, of his boyhood days as a farmer's son in Northern Ireland, when the first cultivating engines were introduced. Robert Eddison, a most likeable gentleman who gave young Allen his chance to get into Fowlers Works to learn the trade, was there at the time. However, by 1926, Mr Allen felt that since his health continued to be so poor, it would be better if he resigned. In that year, he gave up the chairmanship which he had held so commendably since the Association's first meeting in 1915. Although he retired to his 300-acre farm in Northern Ireland, it is sad to say, he was never really well again. His great love for steam, however, did not change in the slightest, because in 1928 we find his son, Major G. W. G. Allen, who had succeeded him as chairman, telling the Annual General Meeting his father resolutely refused to buy a tractor, saying, 'I would not have one of those— something—things on any account'. News of his death in April 1934, was received throughout British steam plough circles with great sadness. He had done a splendid job. Seventy years earlier, the strain of overwork in establishing the steam plough had helped to cause the death of John Fowler. Now it was John Allen who had died earlier than he should have done, because he had burnt up his life in trying to prevent the steam plough engine being superseded by the motor tractor.

The Association carried on till the beginning of the next war in

1939. Some of the discussions at its meetings were about matters such as threatened closure of roads to traction engines, prohibitions on the use of certain bridges, issues of circulars to farmers extolling the virtues of steam work (including an attempt to popularize steam mole draining), wheel damage to the pliable tarred surfaces of roads, compulsion by new Acts of Parliament regarding the use of spark arrestors, the necessity for drivers to be in possession of a driving licence and so on. Throughout these years, however, the membership dwindled as one after another owner died, scrapped or sold his tackle, and went out of business, or as sometimes happened just let the engines stand to rust away; steam ploughs in working order became fewer and fewer.

XVIII

British Built Steam Ploughs at work Around the World

It was only natural that during the second half of the nineteenth century, when Britain was the workshop of the world, her steam ploughs should find a steady market overseas. By 1900 the cultivating tackle trade had so developed, that Fowlers were ready to send a representative to any part of the globe, to give advice to prospective buyers.

In Europe, and more especially in Germany, our steam ploughing sets were soon in favour. The Germans found them particularly useful where sugar beet was farmed on a large scale, and we must remember that it was not until after the sugar shortages of World War I that sugar beet was grown in any quantity as a commercial crop in the British Isles. Quite a few sets also went into France, where one firm of contractors, the Société Anonyme de Labourage à Vapeurs, Soissons, had twelve sets out on hire in 1909. However, France never went in for the steam plough in the same enthusiastic way as Germany.

In Austria and the Balkans, plough engines did some good work, principally in connection with tillage for grain crops. The surplus straw there was often used as fuel in specially adapted straw burning engines. Although steam could be raised as quickly as with coal and maintained quite well, the great drawback was the constant feeding of the straw into the firebox. Moreover, it took 20 cwt. of straw to do the same amount of work as was obtained from 6 cwt. of coal.

The Italians found steam just the thing for the deep ploughing required in vineyard culture. The lava soils in some wine producing areas, presented a tough problem. It was necessary to use single-furrow ploughs set to a depth of thirty inches over terrain on which the bottom layers of almost stone-hard lava peeled up in large solid lumps. As such exposed ground weathered, it soon crumbled down into a fertile earth, ideal for vines. Mussolini, the

148

Italian dictator, is said to have been much impressed when as a boy he saw an English steam threshing engine at work. He retained this interest and when he contemplated draining the huge Pontine Marshes in 1932, he at once suggested steam ploughing tackle for the job. After some tests, in which Fowler's engines appear to have been the only ones to cope really efficiently with the work involved, an order was placed at Leeds for a pair of some of the largest compound ploughing engines every built. They were known as the 'Superba' type, and one of them is illustrated as she stood in the work's yard prior to shipment. Special features of these massive ploughers were 580 yards of $\frac{15}{16}$-inch rope, superheaters and the very high 261 indicated h.p. rating—they were remarkable engines.

The activity on cotton land cultivation in Egypt has already been mentioned. Further south on the African continent much steam plough work was also undertaken. President Kruger, the famous Boer leader, who till the end of his life believed the world was flat, was otherwise sufficiently progressively-minded to become attracted to the idea of developing agriculture in South Africa by use of the steam plough. He even placed an order in 1899 for a set of engines to be used on his own farms. The hard-headed Boer farmers as a whole, however, were not so easily impressed—they reckoned while bullocks could plough at a cost of 1/6d. an acre, the steam plough worked out at 3/6d.

Some Fowler sets found their way to the fabulous De Beers diamond workings around Kimberley. They were employed there on the unique task of levelling and harrowing the immense quantities of diamond bearing blue earth and spoil dug from the Big Hole at Kimberley. About three years ago I received a photograph which shows one of these engines as it now stands preserved in a Kimberley public park commemorating the services rendered in that special earth-levelling job. This is AA No. 12095.

Quite a number of traction engines went out to the British army in the Boer War 1899–1902. A few of them were plough engines, one of which strangely enough was fitted with a dynamo and used temporarily in a small isolated township for emergency lighting. Otherwise, the ploughers were employed on haulage work, or gave other engines which happened to get stuck in sand or river beds a 'haul out' on their cultivating cables.

In the hope of promoting sales in South Africa, Fowlers bought a 15,000-acre farm at Vereeniging. Two sets of engines undertook the cultivation work, and at the same time served as training units for young South African steam ploughing engineers and crews. However, by 1907 hampered by the poor ground and also the drought, the firm had to sell out on this project. It must not be assumed though that Fowlers had a monopoly of the South African trade, because Aveling and Porter secured many orders, as for example the six pairs of compound engines they sold in 1910 to the Natal Government. Besides land work, these Rochester engines were employed with large earth scoops making cuttings and embankments for new railways.

Mr C. A. Smit of Britstown, Cape Province, sent me some interesting details of a pair of Fowler superheated compounds used for many years by the Smartt Syndicate as recollected by his father-in-law, Mr H. Vermaak. It seems that the crew with these engines comprised ten men; two European drivers, one European chargeman, and seven natives for stoking, carting water and fuel, handling the implements and so on. The Europeans were accommodated in a caravan of similar type to that used over here, while the natives lived in a trailer upon which was built a corrugated iron shack. Ploughing, cultivating and levelling for such crops as lucerne, wheat, oats, maize or kaffir beans were the main jobs for which these engines were used. Sometimes, by way of a change, heavy haulage work like moving stone crushers or other ponderous tackle too trying for animals, was undertaken by the ploughers. Once when engaged on a transportation task, one engine was unfortunate enough to get stuck in the mud of a drift, where all twenty-five tons of it sank until only the cylinder top could be seen. Twenty men and a team of mules were occupied for fourteen days carting and fixing iron rails, heavy wooden beams, stones, etc. before this well-bogged engine was got back on to level ground again. Initially, these engines broke up and graded no less than 4,000 acres of virgin land, ploughing and uprooting the native bush in one operation. Previously, using animals, it had been necessary to clear off the bush before any cultivation at all could be attempted. After many years of good service these Fowlers were finally withdrawn in 1936 when it was found that the Munkell tractor was a more economical proposition.

As late as 1952 there were six sets of steam ploughs still at work in the Zambesi District.

Many Fowler sets were put to work in India, often in panther or tiger infested country. One driver, who the first night said he preferred his tent to lodgings in the farm bungalow, soon changed his mind when as he sat by candlelight a panther came prowling and sniffing round the tent flap. Another Fowler driver came back with the tale that on another location while he was introducing ploughing tackle to a farm, no less than five different tigers were seen prowling around.

In parts of India there are areas of clay land which during hot dry weather bake extremely hard, making cultivation very difficult. There are also rich black cotton lands often infested with hariali grass, a most troublesome weed. The coming of the steam plough with its ability to burst open the ground, did much to eradicate this weed, just as it dealt with twitch grass at home. India was one of those countries, too, where it was comparatively easy to train the native population to handle steam engines, and very soon all-Indian crews were in charge of the imported steam plough sets. In Java, plough engines were hindered to some extent when breaking up virgin land by the huge ant heaps found in most wild places.

Japanese agriculturists were not slow to see the advantages they also might gain by introducing steam power. In the islands of the Rising Sun, there is some exceedingly difficult soil called 'Kantenden' found where formerly there were rice fields. When used for rice paddy, these soils could be worked only immediately after rain when they were nothing but mud and sludge. In time, a hard 'pan' or lower floor of hard solidified earth formed about seven inches below the surface, presenting a formidable task when animals were required to break it up for other purposes. The big steam plough engine was a heaven-sent answer for this job, which was performed with deeply set single furrow ploughs. It was also customary on Japanese plantations to plough thirty inches deep as a matter of course and then use a trenching or ridging implement before sowing the sugar cane seed.

Australia is a dry continent, much of it desert, so maybe that is why the thirsty steam plough engine was not imported on a large scale. All the same, some sets went out, especially to Victoria or New South Wales, where one big job of work they did was to break up the wild bush lands in preparation for fruit orchards with twenty-four-inch 'Knifer' implements. In later years, one unusual

Fowler compound set with caterpillar front tracks was exported to Australia, but the idea was not a success.

At first it might not be suspected that the coral reef encircled islands of the Pacific ocean, would be the place in which to find a concentration of cable type steam ploughs. Yet this was in fact so. The engines were a great success in the Hawaaian Islands, breaking up with special heavy five-tined knifer implements, the rough lands when they were first brought under intensive cultivation as western civilization spread across the Pacific. There was some hard soil out there too, as tough as it was anywhere. Often the first thing which had to be done was to put the knifer over the ground to root out great stones which had laid covered since the land first came out of the sea. The crops grown were chiefly sugar cane or pineapples. Culture of the ratoons (young sugar canes) involved very deep work, and the task of rooting out the old pineapple boles made the engines shout out loud. One example of the popularity enjoyed by steam tackle down in the South Seas, may be gained from the fact that in 1923, one Hawaaian sugar plantation company bought its sixteenth set of twin engines. On some of these sugar estates, bagasse, the crushed sugar cane from the mills, was dried and used as fuel in the engines. They then had to have ugly bulbous chimneys to restrict the otherwise continuous shower of sparks. Of course, there was nothing like good old Welsh coal, if that could be afforded over the long sea journey.

South American farmers found work for a large number of British-built tackles. Two Fowler engines sent out in 1884, to the Hacienda Cayalti, Eten, Peru, were still working there quite merrily in 1914. Their splendid record of reliability encouraged the owners to place another order in 1913, when Fowlers shipped compounds Nos. 13633–4 to them to strengthen the team. In 1927, a pair of the older engines was withdrawn from further work, and became outdoor museum pieces. As the flower-bedecked engines were placed one on either side of a main gateway, a band played; and there were drinks all round to the honour of these fine old Fowlers. To my mind, that was a splendid gesture to British-built steam machinery; and I especially like the citation of the occasion, 'Old friends with whom we did not wish to part'. Most homeland owners failed to pay any tribute at all, because usually faithful old plougher warriors either went to scrap, or were left to rust away.

Very little cable-style steam cultivation was done on the mainland of North America. It is true odd sets went from England to Mexico, where there is some of the hardest clay in the world. Prior to the use of steam these soils could not be worked until rain fell and softened them sufficiently for animals to move an implement. At first, both the United States and Canada had a sprinkling of our twin-engine sets just to try them out. They were unimpressed, and turned to their own pet ideas about direct haulage, for which they constructed their own traction engines.

In the West Indian Islands of Cuba and Antiqua, off the western shores of North America, it was a different matter altogether. The steam plough was especially welcome there because the hard dark yellow clay on which sugar cane is grown was extremely difficult to work. In Antiqua, for instance, the usual method of dealing with this tough land was to use the powerful engines in a two-stage system of cultivation. In the first place they worked a plough which piled the clay up into rows of huge lumps. The second operation involved the use of an implement known as a discer which, as its name implies, had rows of sharp steel disc wheels to slice up the clods into small pieces ready for planting the crop. Sometimes, after all the cultivation work was finished, the plough engines were employed on belt work, driving irrigation pumps.

The really important feature about the appearance of these British-built plough engines in various parts of the world, was of course, the fact that they introduced mechanical power into agriculture. The groundwork for the wholesale use of power which we see today, on the fields of the world, was done by steam engines. Most of them were made in England. How proud we ought to be of this revolutionary achievement!

XIX

Some Final Types of Lightweight Steam Engines for Ploughing

The slight advantage of light steam tractors for direct work, which had just managed to tip the scale in their favour at the 1910 Baldock trials, was soon lost to the motor tractors. The big cable engine exponents also had their misgivings about the tractor threat, but on the whole they were so steeped in their proud traditions, that they felt content to rest on their laurels. 'After all,' some said, 'this new motor tractor is a lightweight affair—all right for easy jobs—but nothing would ever replace our powerful compound steam ploughs.' So strongly and loyally was this conviction rooted in the minds of the top people in the trade that when, in the early 1920's, somebody was heard to remark at Fowlers that he thought perhaps the cable engine was on its way out, his remark was scoffed at as a mechanical heresy, 'What! No more steam cable cultivation!—impossible!', they said.

As before, however, the direct tractionists, almost outcasts as they were in the steam game in Britain, acted in a more positive way. They took a careful look at the motor tractor, and saw it was lightness in construction, combined with a high power/weight ratio engine that enabled it to be a really acceptable general purpose machine, which could get on the land in all except the very wettest of weathers. So once again they took up the challenge.

A rather unusual steam machine known as the Summerscales steam tractor was made in 1916 at Keighley, Yorkshire. This was a small three-wheeled tractor with an upright pot boiler and a four-cylinder vee-type poppet valve operated engine rated at 25 i.h.p. The total weight was 3 tons 5 cwt. and only two were built.

Perhaps it would be as well to begin by going back to 1917 when Richard Garrett and Co. of Leiston put on the market their 'Suffolk Punch' light steam tractor. This interesting-looking engine was described by its makers as an agricultural steam

154

ploughing motor. The reason why Suffolk Punch was chosen as a type name, came from the long association of the Garrett family with the breeding of Suffolk Punch horses for farm and other haulage work. This perky little engine with her box-like outline, and comfortably-seated driver placed well forward where he had a good view as he held his motor lorry type steering wheel, was as close as anything in steam ever came to look like a motor-powered tractor. Other special features were the firebox at the front with a stoke hole behind the driver, two compound cylinders with piston valves for steam distribution, a crankshaft and chimney at the rear and a smokebox superheater which gave the steam maximum punching power. In addition a large sized firebox was installed so that low grade fuels could be burnt if the engine were to be used overseas. Her maker's 'write up' said she could pull a four-furrow plough, thresh, and haul a load of ten tons. About a dozen of these steam tractors was built. I saw the one remaining example, now preserved by Mr J. Hutchens at Tricketts Cross, Ferndown, Dorset, when it appeared at the Woburn Abbey and Kilverstone (Norfolk) rallies in the late 1950's. While it is certainly a great novelty, the engine does not give the impression that it has sufficient pulling power to stand up to a motor tractor of equal size. A four-furrow plough on clay land would probably have put her severely to the test. However, it was a notable experiment on the part of the Garrett firm, whose experience with light steam tractions and direct cultivation work was second to none, and if they could not bring out a winner, nobody else had much chance.

In 1924, the Sentinel Wagon Works of Shrewsbury, famous for their steam wagons, introduced two types of light steam tractor. One of these had double gears, steel-tyred wheels and was intended principally for agricultural work. It was claimed it could haul, on average soil, a gang of nine ploughs set nine inches deep. Sentinel Works also turned out what they called a 'Roadless' tractor for direct ploughing. This they built in co-operation with Messrs Roadless Traction of Hounslow. The 'Roadless' feature of these machines was the caterpillar track used in place of hind wheels. Both of these tractors had the usual upright or pot type Sentinel boiler pressed at 275 lb. per square inch, and supplying two-cylinder horizontal simple engines. These engines with cylinders $6\frac{3}{4}'' \times 9''$ were similar to those installed in Sentinel steam wagons. Several of these agricultural tractors went to overseas homes, but little information is available regarding their success on tillage jobs.

Mr A. M. Phillips of Sellershott Green, Letchworth, Herts, designed a light agricultural tractor which was built by the Yorkshire Patent Steam Engine Co. of Hunslet. It came out in 1924 with a compound vertical 50 i.h.p. engine, having oil bath gearing, final chain drive, Ackerman steering and a winding drum fixed on its rear axle. The drum gear enabled a pair of these steam tractors to act as a twin set for cable cultivation duties if required to do so. This 'Yorkshire Power Farmer' (for so it was named) was unusual in appearance, because its horizontal boiler was placed high over the front axle, and it had also an ungainly water tank at the rear. It looks as if it might have made a useful little tractor, but I have no knowledge of its success or otherwise.

Another innovation came from the Sentinel Works in 1927. This time it was an 88 i.h.p. 'Rhinocerus' steam farm tractor intended principally for foreign and colonial markets. The builders claimed that its power/weight ratio was about half that of a compound steam plough engine. The boiler, pressed to 275 lb., was of the usual Super-Sentinel quick steaming type, equipped with a superheater to raise the steam temperature to 600 F. Poppet valves of stainless steel, operated by two camshafts giving long and short 'cut-offs' for running in either direction, supplied steam to the two cylinder double-acting engine. Every attention had been given to the use of light strong metals in order to keep down the weight. It was a tractor type steamer, upon which the latest and best constructional details had been lavished. The final result was to produce a machine, which while she was something like a motor tractor in appearance, still seemed a bit traction-enginish with her crew perched well to the rear. She weighed 10 tons empty and 12 tons when loaded with 10 cwt. of coal and 280 gallons of water. As it was hoped to find a market for this type of engine in dry climates abroad her makers claimed in the advertising brochure that she was so economical she could plough four acres on one tank of water. One of the first of these steam tractors to be built proved herself to be a first-class plougher during trials on Bomford's farms at Pitchill. These 'Rhinocerus' tractors got about the world a bit. One certainly went to Kenya, while others found their way into different parts of South Africa where, no doubt, like the Sentinel steam wagons, they gave a good account of themselves.

Lightweight Steam Engines for Ploughing

Messrs Foden of Sandbach, Cheshire, another firm of steam wagon builders, also tried their hand at a light steam tractor in 1927. This was their 'Foden Roadless' with rear caterpillar tracks. The engine part of this tractor was the same as was used for steam wagons, and had the typical Foden cylinder block. In advertising the virtues of their new steam tractor, Fodens said it could go over uneven ground, mount obstacles, sink into holes, climb steep gradients, cross shallow rivers, enter and leave ponds, swamps and ditches, and generally negotiate terrain which would be impossible to the conventional wheeled pattern tractor. They claimed it could also be used with great advantage in ploughing, especially as there is not the slightest tendency for the tracks to 'pack' the soil. Like the Mann and Yorkshire Engine Co.'s models, this Foden had a large ungainly water tank, telling all the world how unfortunate it was that a steam engine had to lug a burdensome load of water about with it all the time. Later on, in 1931, a spindly feather-weight type of engine appeared as the final Foden farm tractor. Here there was a complete breakaway from tradition, with the bicycle-style spoked wheels and unusual cylinder block perched as if in mid air, over the boiler. How it got on ploughing, I do not know, although it would seem just the thing for a steam-minded market gardener on sandy soil. On rubbers it would have been just the ticket for running around to rallies nowadays.

Bomford Bros, of Evesham, a family long associated with the use and hiring of steam cultivation tackle, were so encouraged by the splendid performance of the Sentinel Rhino that they, too, thought they would see what could be done about making a light-weight engine. What they did was to make a few notes of their main ideas, and without drawings, set about building a pair of Sentinel type engines into a cable ploughing set. They bought, from the Sentinel Works, a pair of Sentinel boilers and engines which had originally been intended for shunting locomotives on the London & North Eastern Railway. In their own repair shops they constructed two chassis to contain the upright Sentinel boilers with engines, and then added cable drums. The pull on the cable was arranged so that it fell at the centre of the chassis, so giving as much transverse resistance as possible, against the tendency of all cable engines to be pulled round at the front to-wards the implement. As this engine was so light, it was all the more essential to do something about counter action against the sideways pull. When finished, these little ploughers were found to be extremely economical, light on repairs and when one of them was tried out in the field sistering a big ploughing engine, it proved it could haul the cultivator equally well while burning less coal. For four seasons this lightweight set did good work, until it was replaced by a caterpillar 30 tractor handled by one man. Mr Douglas Bomford says the boilers would stand up to rough treatment, which seems to have been a remarkable achievement, indicating perhaps that those in favour of the big and unbreakable orthodox engine, might with advantage have turned to something smaller. The twenty-ton compound plougher was a grand, impressive and proud engine, but had probably outgrown itself in hugeness and weight.

Unfortunately, in spite of all the clever design and construction work on the part of those associated with this final batch of lightweight steam engines, the motor tractor gained more and more favour with farmers everywhere.

XX

Steam Ploughing in Germany

The idea of ploughing by steam was, from the very start, given a warm welcome in Germany. Our cousins across the North Sea seemed to share with us the same fascination for well-built, sturdy and dependable two-engine steam tackles. Like us, they thought this way of cultivating land was the answer to growing more food at home to feed a rising industrial population. At first the Germans bought our engines and implements which they found filled the bill very well, but very soon they were thinking of the possibility of building 'Dampfpfluge' (steam plough) engines themselves.

Max Eyth, who was born on May 6, 1836, in a typical German house with high red-tiled roof and dormer windows at Kirchheim, had spent twenty splendid years with John Fowler & Co. at Leeds. It was his return home full of enthusiasm in 1882, which really set things going for a steam plough industry in Germany. By 1890, the first German-built single-engine ploughing tackles had already appeared under trial, near Berlin.

Julius Kemna was an agricultural engineer and steam plough contractor in the town of Breslau. He had bought two sets of Fowler engines for his small business and was highly impressed with their performance, but he improved them with certain of his own modifications. From this success he was encouraged to start thinking about the exciting and patriotic idea of designing his own Kemna all-German type of plough engine. He began by having his Fowler engines kept under close observation; every bad or good point was noted until he had accumulated a mass of valuable information. While a new factory was planned and erected, he contented himself with making a few implements such as balance ploughs for his English engines. The new works were opened in 1894, but Julius died in June 1898 while the prototype plough engine was still in the erecting shop. Fortunately, however, his

159

sons were just as enthusiastic about ploughing by steam and carried on with their father's business at full pressure.

The first Kemna plough engine was finished and left the Breslau works in 1899. It must have been a good one, because other orders quickly came in; there was a particularly big one for engines from the Hapsburg Crown Land Estates in Austria-Hungary. Very soon Kemna were well established as important steam plough builders. By 1905, their trade had grown so much that they opened a second factory in Budapest, followed by a third one, a year or so later, in Prague. Not long afterwards, they claimed to be the most important makers of steam plough engines on the continent of Europe, although in this respect, they probably excluded Fowlers at Leeds, not on the European mainland. In appearance, most Kemna engines looked very much like Fowlers, in fact the resemblance was so striking as to suggest the basic Fowler features were copied. There were, of course, certain variations, such as a single underslung crosshead slide bar, steering chain shaft fixed well forward on the underside of the boiler barrel, and a really satisfactory water pipe coiler at the rear of the coal bunker.

It was not only in Europe that Kemna ploughers were sold in large numbers; they also found ready markets in far away places all over the world. Prior to 1914, they were being exported to Argentina, Peru, Puerto-Rico, India, and the German East African Colonies, as well as to Asia Minor. This wide sales area only goes to show how far the Germans had gone towards making themselves formidable competitors with the British in supplying the world's demand for cable type plough engines.

Whilst dealing with steam work in Germany, it seems only right to say something about superheated steam, because the Germans were always very keen on it. If such a fact may be mentioned in this book about the glory of steam, it was while Dr Rudolph Diesel, the German scientist, was experimenting in 1888 with superheated steam, that he stumbled on the idea of the solid fuel injection internal combustion engine, now named after him. As early as 1900, the Kemna firm was experimenting with superheated steam in their plough engines, but due to the imperfect apparatus then available, they did not meet with success. In 1907, however, the prospects improved when Dr William Schmidt invented his successful smoke tube superheater at the big railway locomotive building centre of Cassel. The basic principles of

17. One of J. B. Carr's Fowlers takes time off from dredging to
pull up some trees

[*Photo:* W. Dennis Moss, Esq.

Sale of John Patten's steam ploughing equipment, Little
Hadham, Herts., 1960

18. 1918 Fowler 'BB' No. 15226 'Tiny Tim'. The cultivator is approaching from engine seen in the centre on the opposite side of the field. Working 1960 for Taylor Bros., Wimbish, Essex

[*Photo:* Barry J. Finch

Steam ploughing: Driver's view of Fowler six-furrow anti-balance plough at work at Chrishall, Royston, Herts., in October, 1950

[*Photo:* F. C. L. Stokes, Esq.

Schmidt's findings were that if the wet steam normally used were given additional heat after it left the boiler, the result was a great improvement in its working characteristics. This extra heat was imparted to the steam on its journey between regulator valve and steam chest by passing it through a series of pipe coils exposed to the hot gases of the fire. These pipes could be either inside the smoke tubes or in the smokebox only, and it was this separate pipe system which was known as the superheater. As the steam passed through the superheater, droplets of moisture were dried out and evaporated to give a larger volume of steam, which due to its increased heat content, was also nearer a true gas and very resistant to condensation. This meant that whereas wet steam was quick to condense in the cylinders, superheated steam had the opposite quality. It hung on to its heat and a whiff of it acted like a slowly exploding bomb in front of the pistons, and it kept on pushing hard. There was as much difference between wet and superheated steam, as there was between a half-baked and a well-baked cake. As soon as it was plain that Schmidt's superheater was a success on railway engines, Kemna decided their plough engines should have the advantage of superheated steam. They redesigned their earlier apparatus and from 1908 onwards a new type of superheater was fitted to Kemna steam ploughs. Shortly afterwards it became almost standard practice in Germany to include a superheater.

In Britain superheating on steam ploughs never gained any popularity, although it was accepted as one of the most outstanding improvements made on the railway locomotive since George Stephenson's Rocket. I well know from my own experience as a fireman on the old L.N.E.R. that a superheated locomotive was streets ahead of a wet steamer. Although Fowlers appear to have been barred from using Schmidt's patents in Britain, they did try superheating several times. The engines so fitted were piston valve compounds. More often than not, however, it was not long before the purchasers of the engines removed the smokebox superheater fittings. The main trouble seems to have been that the very high steam temperatures of around 600° F. called for the best of cylinder lubricants, otherwise the valve faces soon wore down. The piston and valve rod packings did not stand up well either to the burning temperatures of the steam. One old East Anglian steam plough driver to whom I once talked at a North Weald rally, told me his experience of superheated engines was that the farmers liked them because they were more economical on fuel and water; but it was the owner's repair bills for cut valves and so on, that put superheating out of favour.

Another argument against the superheated engines was that English fields were too small and the bouts too short to enable the engines to obtain full benefit from superheated steam. Success with superheating depends upon holding the high steam temperature for longish periods. However, Schmidt and his superheater ultimately gained world-wide fame, even though his idea was not acceptable in British steam plough circles.

The firm of Borsig of Berlin-Tagel, grew to be the steam plough builders of second importance in Germany. Their engines were often built on the Ventski principle, a design which incorporated such special features as an unstayed cylindrical firebox and a long smokebox upon which the cylinder casting was bolted. With the cylinder in this position there was no need to make the holding down studs steam-tight, and there was less risk of water getting into the cylinders whenever engines happened to have their heads down on a steep hillside. There was also an arrangement for fixing the crankshaft side brackets, so that the heat from the boiler was not conducted up to the motion bearings. There may have been something in this, since boiler heat can make the shafts of Aveling & Porter and Fowler steam rollers almost too hot to bear one's hand on them. The Borsig boilers were pressed to 15 atmospheres (220 lb. per sq. in.) and their post-war engines revolved fast at 384 r.p.m. In the first decade following the war, not only Borsig but other German makers continued to give far more thought and attention to improvements in design, than was the case in England. For instance, in 1927, Borsig designers were still so assured in their own minds about the future of steam, that they built a new type single cylinder 60–80 h.p. engine intended for farmers who had limited acreages. Among other experiments was the Borsig model with cylinders and rope drum at the rear end, following the former example of Howard's English 'Farmer's Friend' engine, although this German engine differed in that its cylinders were arranged vertically.

In Germany, there were extensive areas of soft moorland wilderness never before cultivated on any but a small scale. In order to bring these under the sway of steam, both Kemna and Borsig devised particular types of engines which had very wide and smooth rims on both front and rear wheels, these engines were very steamrollerish in appearance. The purpose of the extra wheel width was, of course, to give the engines more support on the softish moorlands and prevent them from sinking in as ordinary wheels might. It is true Fowlers had for sale special extension rims for the hind wheels of their engines when used on Fenland work, but the German method was to have really wide wheels all round. As a matter of general interest, a census taken in 1925 showed there were still 1,600 steam plough sets left in Germany, compared with 7,000 motor tractors.

Other German steam plough makers were A. Heucke of Magdeburg; Rheinmetall of Düsseldorf; Maschinenbau-Gesellschaft Heilbronn a Nekar and R. Wolf of Magdeburg. The great point of interest about German ploughing engine tackle construction is that it was Germany alone, with its *Zweimaschinen dampfpluge*, which followed the British lead and pattern of cable work with two engines. They liked it, and went in for it as enthusiastically as we did. So far as I know, none of these German ploughers were imported into the British Isles, although at least one of their traction engines found its way here as part of a reparations payment after the First World War. This engine worked in the Isle of Man until 1959 when she was laid aside and I intended to buy her for preservation. I was told that she had been a good engine, except for a tendency to fuse the lead plug when running down steep hills. She was fitted with a set of governors said to be remarkable for their sensitive control of speed, far better than any known make of English governors. Unfortunately, the boiler inspector had on his last inspection requested that the cast iron superheater header should be changed for a steel one, and since this looked like being a biggish job, I had to go back on my intention to buy her. Shortly afterwards, she was cut up in the Isle of Man by an English scrap metal firm, which had bought a number of other engines on the island. The scrap metal firm went bankrupt a little later and I lost my £10 deposit into the bargain. The liquidator, when advising the creditors that no dividend whatsoever would be payable, said the scrap iron from the engines had disappeared and the Manxman appointed to look after the yard had died!

It may be of interest to mention that Fowlers had a factory at Magdeburg for their continental trade. When war broke out in

1914, the German government took over these works until March 1917 when they sold them to the firm of R. Wolf. After the war there was a lengthy litigation over the matter, which eventually ended rather badly for Fowlers, who were awarded only about one-third of the value of the installations.

The Germans, like the Italians and more especially the French, tried their hand at electrically-powered cable cultivating machinery. One drawback was the need to trail out a long cable to each traction unit, and this apparently had its dangers too because in France at least several men were electrocuted. Such a way of working may, of course, become practical in the future. The motor tractor will not keep its popularity for ever.

Further confirmation of the way the Germans hung on to steam may be gathered from mention made by the Rev. Philip A. Wright, M.B.E., of Woodford Bridge, Essex, when writing in his Chairman's column of the April 1963 issue of *Steaming* (the National Traction Engine Club's quarterly journal) that a correspondent in Germany had recently written to him saying that one contractor had sixteen steam ploughs, of 1953 construction, still at work.

In conclusion, I would like to say how the Germans have paid a touching tribute to Max Eyth, the father of their steam cultivation activities. He died on August 25, 1906; but he is still honoured by the preservation of the house in which he was born, as well as by a low stone memorial erected June 1933 in a public park at Ulm. This example ought really to inspire us into doing something similar in commemoration of the great work done by John Fowler.

XXI

The Last Years of Commercial Activity with Steam Ploughs in Great Britain

A great deal has been said already about the big cultivation job done by the British steam ploughing contractors and owners during the 1914–18 war. It would not be right though to assume all the good work was connected with land tillage, because other important tasks were successfully accomplished. There was, for instance, the cutting of 2-ft deep sewage trenches on the hillsides of the army camp sites at Clipstone and Cannock Chase by ploughing engines belonging to J. Mather & Co., Newark, Notts. There is, too, a photograph in the War Museum showing a Fowler compound plougher, fitted with a cab, standing on the beaches at Gallipoli. What did she do out there? One likely task may have been to use her cable to haul in loaded lighters from ships which had to anchor some distance out because of the shallow inshore seas. I doubt, too, when the troops were eventually evacuated in secret from that ill-fated expedition (which had failed to drive a wedge between the Germans and the Turks), whether they even bothered to try to load her up again. It was more likely that a charge of dynamite disabled her for ever.

Fowlers did some demonstrations for the Russians to show what a pair of compound plough engines could do in the way of cutting trenches for their troops, although it is doubtful whether the Tzar afterwards placed any orders for engines to do this work on the fighting fronts. Dick Bateman of West Drayton, who drove traction engines with the British army in France, says plough engines were also tried there for trench cutting behind the Western Front. The smoke from the engines, however, made them too obvious targets for German aircraft or artillery, and the idea had to be abandoned.

In 1923, Mrs Susannah Kennelly, a Peterborough newspaper seller, came into a fortune of £10,000. Strange as it may seem to some, her first move was to place an order with John Fowler for a complete two-engine set of ploughing tackle. Her husband, an agricultural mechanic, proposed to work the engines himself. What actually happened to this venture I have not been able to discover. Thirty years later, I enquired casually of an old lady stall keeper in Peterborough Open Market, whether she knew anything of Mrs Kennelly's creditable investment in steam. After a moment's bemused contemplation she replied, 'Never heard of it—all I can say is she wanted her head testing'.

In the museum at Peterborough there are one or two exhibits of model steam cable implements made by a local firm, Barford and Perkins, who did a good bit of implement manufacture without going so far as to build the actual engines. While looking at these one Saturday, I chanced to get into conversation with one of the attendants. He told me that years previously he had been a fisherman in the Wash. As the fishing vessels cruised homewards, one of the first indications the crews got of their landfall in clear weather was the smudge of smoke sent up by a set of steam ploughs owned by a man near Sutton Bridge in Lincolnshire. This smoke signal could be seen long before the first buoy was sighted. I suppose this was a unique instance of steam ploughs serving as a navigational aid.

The dislike of other road users for traction engines never abated. Right from the beginning the opposition had been nasty, particularly from people interested in the use of horses, animals notorious for the ready way in which they shied or 'ran away' if required to pass a steam engine on the road. There are many accounts of trouble with horse owners, but perhaps one early story will be sufficient. In May 1880, Daniel Wright, described as an engineer of Fazely, near Birmingham, was summoned by the police for not stopping the plough engine he was driving when a horse became restive. The groom in charge of a horse and brougham had signalled Wright to stop, but apparently he decided not to do so. The lady passenger in the vehicle was so alarmed at the horse's fright that she jumped out just before the animal bolted with the vehicle into a field gateway. The magistrate ruled that a boy ought to have been 60 yards in front of the engine to give warning because as the law stood, an engine driver was completely in the hands of anyone in charge of a horse. It was only necessary for any horseman to lift his hand to oblige the driver of a steam engine on the road to stop at once. Wright was fined £5 and costs—rather a stiff sentence for a poor engine driver probably earning less than £1 a week.

After the First World War, when motors began to displace horses, motorists were just as awkward about the traction engine, which they considered damaged their roads and often got in their way. Steam traction was always regarded as public enemy No. 1 on the highways of Britain. The country policemen never missed a chance to 'catch' a steam engine driver if they could possibly do so—immediately an engine went out on the road, the police seemed to regard it as dangerous as a lion escaped from a zoo.

A more serious attack came from the County Councils, who in 1923, threatened the closure of many supposedly weak bridges. Admittedly, the big A/A and B/B type compound engines were very heavy and their 20-ton weight was certainly worthy of consideration by any council.

There were clauses in the 1894 Roads Act and 1898 Locomotive Act enabling Local Authorities to prohibit the passage of heavy vehicles over uncertain bridges, and attention had been called to the dangers of collapse by the failure of an 1842 bridge at Penybout, Radnor, when a steam tractor with living van and a water cart had gone headlong into the river below and the driver saved himself by swimming away. Apart from the bridges now under the jurisdiction of the Councils, there were also in 1923 no less than sixty-four private toll roads and bridges in England and Wales, and it was easy for the owners of these to refuse access to steam ploughs on the move. The proposed closure of a bridge in one particular district meant that in future steam plough sets moving between one farm and another would have to make a detour of no less than seventeen miles, instead of travelling three miles over the bridge by the direct route. All through the twenties and thirties, the Steam Cultivation Development Association was plagued with these threatened closures of roads or bridges. The size of the problem may be gauged by the announced intention of Wiltshire County Council alone to bar traction engines from using thirty-seven of their bridges. Railway and Canal companies were also most un-cooperative, contending that quite strong bridge structures were incapable of carrying plough engines. Now bridges have queer characteristics in their weight-carrying abilities as was shown during some tests carried out in 1936. A bridge at Croft near Derby, on which the owners proposed a 5-ton load limit, began to crack under a 40-ton load, but did not finally fail until 80 tons were placed upon it. Another red brick bridge, with a 3-ton axle limit, or total vehicle weight of 8 tons, tested on the Yardley Wood Road near Birmingham, withstood an 85-ton centrally-placed load with a mere $\frac{1}{4}$ inch deflection and held $126\frac{1}{2}$ tons before failure. However, this was only one side of the picture because some bridges which looked perfectly safe and strong, let ploughing engines go through straight into streams below! Messrs Beeby Bros of Rempstone, Leics., had a case where one of their single-cylinder plough engines took the whole crown of the bridge down with her. So you see there was some excuse for the attitude of the Council Surveyors in trying to safeguard themselves again future criticisms should accidents like that ever happen in their particular areas.

As more and more of the roads of Britain were covered with tarred surfaces, a new uproar arose over the tendency of ploughing engine hind wheels to pick up the tar when it was soft on a hot summer's day. Even if they did not badly indent or lift off the surfaces of the roads, the straked rear wheels usually made deep impressions in the tarred surface whenever the heavy engines, which had no differential compensation, tried to man-oeuvre into field gateways. One of the more notable instances of road damage trouble which the S.C.D.A. had to cope with concerned a set of single cylinder ploughing engines at Eastbourne. These engines belonged to Alfred Fuller and Sons, local ploughing contractors, upon whose business the engines were being moved on July 14, 1926. It was an extremely hot day and the County Borough of Eastbourne claimed they did £225 worth of damage on the East Dean Road of the Eastbourne–Brighton highway. This road had recently been given a first-class surface and on the day in question, as the two engines moved towards Eastbourne between one and two in the afternoon, they peeled off the new facing in an alarming fashion. What happened was afterwards described by one driver to the judge, 'The road picked up like pudding'. Each hind wheel left a trail of bared foundation with the new top littered in sticky-looking lumps all over the place. As soon as the drivers saw what was happening, the engines were stopped.

Shocking as the damage already was, the men decided, since they could not at that point draw the tackle off the roadway, there was nothing for it but to see if the road a little farther ahead held them any better. After another 200 yards progress however, it was clear to Mr Fuller's son (who was driving one engine) that they must at all costs get off the road as soon as possible. The first engine was then unhooked from the cultivator and living van which it was hauling, and taken on a short distance to where there was a side turning into the Beachy Head Road. As soon as this engine had turned into the side road, its rope was let out and the load it had left behind was hauled forward. But since the shedding of the load had in no way lessened the tendency of the iron-straked driving wheels to pick up the tar the second engine was allowed to follow with plough and water cart. As the men looked back, the road really did look dreadful, but they felt they had done all they could to keep the damage to a minimum.

When members of the Steam Cultivation Development Association heard of all this and the impending prosecution of Mr Fuller, they immediately rallied to arms. They knew very well that if this case were lost, it would be the thin end of the wedge to prise the steam plough off the king's highways entirely. Mr Fuller, beset by bad trade, was unable to pay anything much himself towards a County Court defence of £500 or possibly £1,500 if the case went to the High Court. Captain Allen and Mr Goode generously offered £50 apiece, Mr Grundy £20 and Mr Wilder £10 to start a fighting fund. The case eventually did go for a two-day hearing during October 1928, before Mr Justice McCardie, in the High Court of Justice in London. The plaintiffs were the Mayor, Alderman and Burgesses of the County Borough of Eastbourne, suing Mr Fuller, the defendant, for £255 damage to their road. Relying upon the Highways and Locomotives Act 1878, as amended by the Locomotive Act 1898, it was submitted for the plaintiffs that the movement of the two ploughing engines was extraordinary traffic and thereby the engines became a nuisance. Fortunately, the S.C.D.A. had briefed a good defence counsel who proved to the judge's satisfaction these cultivating engines had worked in the neighbourhood for many years and several farmers still needed them, and in these circumstances there was nothing extraordinary about their journey. Further, by a clever argument, it was established that the new road surface in question, made of mexphalte and granite chips 3 inches deep, was not uniformly laid as specified in the Borough's contract for construction. There were variations in the amount of mexphalte and granite chips at various places, so the road was not precisely what it was claimed to be. Summing up, the judge began by saying the circumstances of the case were most unusual. He then went on to quote several previous judgements for road damage, before dismissing the charge with costs. All the old steam operators were delighted with this verdict and went home feeling as though they ought to ring the bells in their village church towers.

The point which most likely defeated the plaintiffs was that the engines were ruled to be 'ordinary' traffic as they moved from farm to farm, about the business of land cultivation. To have succeeded in the claim for road damage, it would have been necessary for the Borough authority to prove that the plough engines' journey on the road was of an 'extraordinary' nature. The Acts protected the roads' authorities from having to make special provisions in the strength of their highways in such circumstances. Today, should a heavy iron-shod engine in the hands of an enthusiast do similar damage to a soft surface, while just out for a spin, or on the way to a rally, there appears every likelihood a prosecution would succeed on the grounds that the traffic was 'extraordinary'. In order to avoid any further irritation of local authorities, the S.C.D.A. wisely advised all its members to endeavour as much as possible to move tackle either early in the morning or at the cool of evening, during hot summer weather.

It must not be assumed that only defensive action was taken when the tractor was seriously challenging the steam engine. On the contrary, the old stalwarts of the S.C.D.A., or sometimes individuals, never lost an opportunity to show what steam could do. As early as the autumn of 1918, a very interesting competition had taken place. Joseph Bowser, J.P., chairman of the Boston and District Agricultural Society, wagered £100 with Messrs Dennis and Sons Ltd, farmers of Kirton that he could plough ten acres with two Whiting Bull tractors as well as their traction engines and at less cost. The stake money was to be given to the local Farmers' Red Cross Fund. Dennis's Fowler compounds won. They did the job at a cost of 5/9¼d. an acre, against the tractors' 6/2½d.

On an average you could say a set of compounds such as these would burn about 1¾ cwt. of coal per acre when ploughing. Mr

Young, the judge, said he had to admit the tractors did very well, but they had not had such experienced men on them as the steam ploughs. Messrs Dennis were long established owners and users of steam tackle whose crews were seasoned, well trained and tip-top chaps, as good as any in Lincolnshire. At a tractor demonstration held in 1922, on the Duke of Marlborough's estate, a set of compound engines arrived unexpectedly. Working under controlled test conditions, with coal weighed on and land carefully measured, it was found the steamers ploughed very nicely on 95 lb. of coal each per acre, and their fuel costs were lower than those of the tractors. The good old steam plough was still supreme.

Mole draining was another activity upon which many hopes were placed. It was quite true that while the tractor was at first able to steal the lighter cultivating and ploughing work from the heavy steam tackle it could not yet do the extremely hard work of mole draining. John Fowler's early mole drainer has been mentioned, but it will not be out of place to say a word or two again by way of explanation. With mole draining the object is to force through the ground, at a depth of about twenty to twenty-four inches, a bullet-shaped plug or snout, in order to leave a 'squeezed out' hole in clay soil. This type of pipeless drainage cannot be made in sand, because the hole fills in again immediately, and even in clay, it is better done in wet weather when the earth is soft enough to allow the compressing action to take place. About $5\frac{1}{2}$ yards are allowed between each 'mole' and the work is more often than not undertaken on permanent grassland. The engines work in pairs as for cultivation, but only one of them does the pull for the 'working' run. The other engine merely has its rope out to draw the mole drainer back empty across the field. As the pulling is as hard a job as any which steam plough engines ever encountered, the rope from the 'load pulling' engine was run out to a large horizontal sheave wheel on the drainer, and then back to the engine where it was lashed fast to the rear wheel. By this arrangement a 'double' purchase' pull was given to the implement, which could not have been hauled by direct winding alone. Another interesting point about mole drainers is that the steersman had to stand up on the implement all the time, no seat was provided. The 4-inch diameter mole holes were, of course, a compromise. They were used instead of laying the more expensive piped drains, and were effective for a period of about twenty years. If you imagine the result of a big bullet fired through a lump of cheese, that closely resembles the sort of hole a mole drainer leaves in clay.

I well recall once in the summer of 1928 whilst out for an evening stroll with several engine-cleaning pals from the Locomotive Depot at Hatfield, that we came across a set of compounds set up for mole draining, but, unfortunately, the men had packed up for the day. We did, however, get through the tall hawthorn hedge bordering the field near Whitegates, now swallowed up by De Havilland's airfield, to have a closer look at the engines. After satisfying the farmer who appeared suddenly, and at first a bit threatening, that all we wanted was to have a look round, he joined in by describing one or two points to us. Even though, in a general way, some few extra mole draining jobs were secured for the steam tackles, these were insufficient to make up for the lost cultivation work. Later on, one Bedfordshire owner tried to cheapen the steam side by using a plough engine for the pulling, and a tractor for hauling back the drainer on the 'empty' runs. Mole draining prices varied a good deal; in 1930 Mr Wilder of Wallingford was charging 28/- an acre, although in other districts the work was undertaken for half that rate.

The contractors still cut prices to get work, and so find a full week's employment for the steam plough gangs.

When steam work began to go out of fashion in the early twenties, one of the first effects noticed was the dropping off in orders for new sets. All the steam plough firms felt the pinch and in 1921, Aveling and Porter stopped building plough engines altogether. Fowlers and McLarens carried on with the few jobs on their order books, while at the same time they both turned an eye to the possibilities of cable sets powered by internal combustion engines. What dreadfully plain and uninspiring machines they looked! They were boxes on wheels compared with the upstanding, graceful and magnificent steam compounds. These oiler cable tackles never looked right in any way, so it is not surprising they did not succeed. Meanwhile, both McLarens and Fowlers kept as good a heart as they could by exhibiting their steam ploughs at the usual shows and exhibitions. There was a Fowler compound at the British Empire Exhibition 1924-25 at Wembley, and those who attended the December 1925 Smithfield Show saw a lovely Fowler B/B1 class compound on show. She worked at 200 lb. boiler pressure, had 650 yards of rope, two injectors and 26 feet of suction hose. Two driving pins were provided in her hind wheels in order to give a more solid propulsion, and she also had the luxury of a ferodo band handbrake operating on the driving plate of one rear wheel. What a splendid engine she must have looked on her stand! In the early days Fowlers used to paint all their engines black, but after 1920 some ploughing engines were turned out in a brownish maroon. Black looked fine on a plougher, especially when set off to the best advantage by lining out in three colours. There was a broad orange-brown band edged with a thin red vermilion line, and if room permitted an inch inside this was a $\frac{1}{16}$" chrome yellow line. The big ends were always a striking vermilion red. The firm's paintwork was well done. I have been shown some 'Fowler's paint' by a proud owner of a 45-year-old BB compound.

The last Fowler set of plough engines sold to an English customer went to Nottingham Corporation in 1933. The final pair of all were made in 1935 and shipped to Formosa. So ended a wonderful era of plough engine building—the finest record of its kind in the world.

New laws on licensing and use of the roads were another headache for the contractors. The 1930 Road Traffic Act introduced the obligation that every mechanically propelled vehicle (except a steam engine registered before November 1, 1933 and capable of being reversed) should have a brake. This did not cause much bother, because all the plough engines could be reversed and so obtain braking power. For the first time, however, drivers and steersmen were required to be twenty-one or over and in possession of licences. Chaps who had driven engines all their lives now had to get a licence, just like an ordinary motorist. Fortunately, the Ministry of Transport driving test examiners did not in all instances know very much about how to drive a 20-ton plougher. Old Percy Grundy used to tell the tale how his men were passed. The examiner called at his Glendon Works, Kettering. After having a look at an engine in the yard, he asked if one of the men could drive it round the adjoining streets. Mr Grundy obliged; and on being asked whether all his men could drive as well, replied that they could. The examiner then issued driving licences to the lot. Nobody liked the part about licensed steersmen, so several groups of traction engine owners lobbied their M.P.s with sufficient pressure to get some mitigation in the 1936

Road Traffic Act. From then on steersmen could be employed without licences, on the direct understanding they were always under the control of a licensed driver. Mr Grundy soon got caught with one of his drivers. As he was short of men, he had been only too glad to employ a young driver of 18, 'A good chap too', he said he was. It was all right while the lad drove in the fields; but on the very day that he took a chance when moving to another farm, the local policeman was watching and promptly 'nabbed' him. The magistrate imposed a fine of £5. While mentioning this instance of attention by the police, another story about a set of Allen engines in Buckinghamshire might be told. The engines had stopped to pick up water from a river. Traffic passed all right but up came a 'Bobby' and said the engines were causing an obstruction. When the foreman, in giving evidence, told the magistrate, he just did not understand why they were summoned, the policeman said, 'It was Newbury Race Day'! Luckily, the case had been brought under a 1904 Act which applied only to motor cars; and in this instance the steam plough-men got off scot-free.

One of the larger contracting firms to sell out quite early was T. B. Kitchener and Co., Steam Plough Works, of Potton, Bedford-shire. Their nine sets of engines were all called in from work and auctioned on September 7, 1927, on the voluntary liquidation of the firm. There was one set of 8 n.h.p. singles fitted with Pickering governors, five sets of 14 n.h.p. singles, one odd spare single engine and three fine sets of 16 n.h.p. BB compounds. It was rather interesting to find that most of the single engines had been re-boilered by the Oxford Steam Plough Co. and, besides, several of the cultivators had improved axles fitted by the same firm. The compounds had rear wheels six feet in diameter with treads 22 inches wide. A thousand or so people attended this sale, which proved a very good one. A 40-year-old set of what were described as 'Cog' engines fetched £50 and the best set of compounds—Nos. 15214/5—made £1,500. By 1930, from the farmers' business-minded viewpoint, steam ploughing had become expensive. At one meeting of the Association, Mr Wilder said, 'I cannot help thinking that the steam plough, even if you have the men, which some of us have not today, is not an economical proposition'. He went on to show that a one-man operated caterpillar tractor using a six-furrow plough could do 10 acres in a 12-hour day. The best any set of steam tackle could achieve was about 12 acres, but with a crew of four men and a boy to pay. The same doleful tale was repeated by Mr Molesworth of Ketton near Stamford: 'What with labour shortages, no horses left on the farms to cart the water, the fact the old steam hands are about extinct, and younger men will not rough it as they used to—it is wellnigh impossible to continue.' As one looks back, one feels more and more admiration for those fine old steam owners who stuck it out so long trying to keep their engines at work. They had steam in their blood all right.

Ward and Dale of Sleaford, who at one time had twenty-five sets out in the fields and one spare set in the yard, received a bit of a shaking when Mr Wallhead, their manager, died in May 1937. He was the representative for Lincolnshire on the S.C.D.A. committee where he had served since 1917, and was regarded as a good practical man, an old friend and colleague. By chance I met him once in the summer of 1935. I was on my way to spend the weekend with my parents, and called in Whipples' garage in Grantham to see Bill Millar, an old school pal with whom I had shared boyish steam plough experiences. By luck Mr Wallhead came into the garage to collect the firm's 1929 Austin saloon to drive home to Sleaford. Bill introduced me and I was offered a lift out to Barkston. We chatted about steam ploughing and he told me the old foreman Dick Stephenson had retired long since and died while living with a married daughter in Sheffield. The steam contracting at Sleaford wasn't going too well and they were obliged to charge as little as 10s an acre for cultivating, to compete with tractors.

One evening while out for a run in my 1926 Clyno Car, during the late summer of 1937, I came across a set of Ward and Dale compounds doing a bit of interesting work quite close to Crowland Abbey, Lincs. The engines were in a field where the fen soil was as black and almost as soft as soot. They were working with a two-furrow deeply-set plough. The land lay below the level of the Welland embankment, down the rough grassed slopes of which I had to scramble in order to get to one engine. The job was a tricky one for the drivers because, as with many fenland fields, there were hidden trunks of ancient oaks a foot or so below the surface. Every now and again the ploughshares met a piece of this bog oak timber. Each driver kept his hand on the regulator ready to shut off immediately the ploughman signalled 'stop' by putting up both hands. More often, though, the concentrated and practised eye of the driver, saw his rope 'snatch' first or felt the extra strain on the engine as the plough either 'glanced' or stuck on a buried tree. As a matter of interest, those old oaks lie on the clay floor of the Fenland, and are covered with deposits of black soils left behind by the meres after the flood. Nobody knows just when the trees fell, but men on certain Fenland farms say they all lie 'South west', so perhaps one great wind-storm levelled them in prehistoric days.

By 1938 it had become difficult for Ward and Dale to keep in business, and they were actually losing money. That year six sets of the valiant old singles were cut up for scrap. Their dismembered rods, wheels and general remains, together with the broken up implements, filled fifty railway wagons in Sleaford Yard. They say some of this scrap metal went to Hitler's Germany, and I did hear it said by a farm hand, three years later, when German aircraft began their attacks on this country, that, 'As likely as not some of the bombs were made of Ward and Dale's old engines'.

Early in November 1938, a set of Ward and Dale compounds which had been working south-west of Grantham, called at my old village of Barkston on their way home for the winter. They did about a week's work on some clay and ironstone land on what was commonly known as 'Up the Little Hill'. The twenty or so acres they cultivated was left in a bit of a mess, for the ground was too wet and the implement 'bunged' at least once each pull. They got on better with ploughing barley stubble and my father, who was then retired with time on his hands, went to have a look at the engines while they did the ploughing. In his opinion, 'They had made a good job of that'. On the Sunday afternoon, the cookboy arrived from Sleaford on his motor cycle to light the engine fires in readiness for an early start on Monday morning. He filled the fireboxes with wheat straw, and as soon as this was well alight, some lumps of coal were thrown in after it. The heat given out from burning straw is quite remarkable, and engine fires so lit, burn slowly all night without any risk of raising more than a few pounds of steam by morning. Of course, a damper plate was laid over the tops of the chimneys, and the ashpan dampers were closed. That was the last job done by any of Ward and Dale's ploughing engines in our parish, and perhaps the very last any of their engines ever did while in their ownership.

Last Years of Commercial Activity with Steam Ploughs

The firm sold out on March 10, 1939, lock, stock and barrel. That day, ten sets of their compounds, all that remained of the once largest steam cultivating concern in the world, were auctioned by Escritt and Barrell of Grantham. The auctioneer began by saying it was the unreserved dispersal sale necessary on the closing down of the well-known and old established steam ploughing and cultivating contractors, Ward and Dale. He went on to point out the engines were in excellent working order, having been recently overhauled. The seventeen double-boarded living vans would be ideal for seaside use, etc. These vans he said would be sold fully equipped with cooking stove, store boxes, cupboards, cooking utensils, cutlery and crockery, as well as the usual bedding and linen for five men. Although with the passing of time, lists and details of prices lose much of their interest and significance, perhaps an exception ought to be made in this particular instance. The engines sold that day were as follows:

Pairs of Engines

H.P.	Type	Nos.	Sale price for pair of engines		Purchaser
			£	s	
18	AA	15256/7	110	0	Mr Cole
16	BB	15148/9	68	0	Hempsall, Redmile
16	BB	15196/7	87	0	H. Roads
16	BB	15412/3	80	0	Shepperson
16	BB	15204/5	88	0	Chivers and Sons, Histon
16	BB	15346/7	72	10	Stamp, Market Rasen
16	BB	15230/1	58	0	?
16	BB	15194/5	104	0	?
16	BB	15154/5	140	0	H. C. Burgess, Witham-on-the-Hill
16*	BBS	13978/9	68	0	Crawford, Frithville

* Fitted with piston valves and two-speed gear.

The average prices for implements and vans were:

Ploughs	£8	Water carts	£15
Cultivators	£12	Living vans	£30

This was not a good sale for Ward and Dale. Really good engines went for a song—at hardly more than scrap prices. 'Mareham House', the manager's residence, the yard, office and all equipment were sold that day. One evening in 1940, I happened to have half an hour between trains at Sleaford so I popped round to see what changes had taken place. I found the army occupying the buildings. The firm's brass plate engraved to show their trade as steam cultivating contractors was, however, still on the wall at the entrance to the offices. It has since disappeared.

A condition of the winding-up sale was that, 'All lots to be absolutely removed before 6th April 1939'. That was not the case, because at least one engine still stands in the little field round the corner from the main depot (formerly used as an out-parking or winter standing ground) where she was sold. It almost seems as if she has silently defied eviction. She stands there with a fair-sized sapling grown up between one wheel and the firebox, as a rust-ridden memorial to the former great and busy days of Ward and Dale. Keeping forlorn company with her are seven other plough engines and a collection of implements, living vans and water carts, all owned by Mr H. Roads of Caxton, Cambridgeshire. He was a buyer at the sale and has since added to his collection considerably. He began saving plough engines from the scrap man long before the present interest in traction engines. This stands well to his credit. When I chanced to come across him in the spring of 1961 while he was planting potatoes in one of his fields in which four ploughing engines are stored in a corner below a little wood at Orwell, Cambs., he told me he has no less than ten sets saved in various places. The reason seemed plain when he just said, 'I love them', and no doubt he does, too.

When the Second World War broke out in 1939, a government order went round to count the steam plough sets still in working order. This seems to have been in the nature of a 'Just in case' enquiry, however, because very few sets were put back to work on the land. By 1940 there was a general air of despondency in the debates of the S.C.D.A., and many members began to see the time would soon come when they would have to wind up altogether. That fine old steam man, Mr Briggs of Stamford, stood up before them and challenged, 'What beats steam?', to which Captain Allen replied—'£.s.d.'.

At the Association's meeting in July 1941, it was decided to amalgamate with the National Traction Engine Owners and Users' Association, an older organization which had concerned itself mainly with threshing and haulage activities. One of the old members, Mr Reynolds of Bedford, when asked by letter earlier to give his approval to the proposed amalgamation, said he was now too old to have the worry of such a decision and would have to leave it to his sons to decide. Percy Grundy was in the chair at the last meeting. This is what he said, 'It is sad to think this will probably be the last meeting we shall ever have, but as the times alter, one must expect this kind of thing'. He himself carried on with his steam sets at Kettering where they were known familiarly as 'Tommy's', 'Nips', or 'Bozzy's' after the foremen who were in charge of them. Some of Grundy's sets did big dredging jobs, as for example, making the lake at Wickstead Park, and the lake at Petersfield in Hampshire, and they also completed a long contract for dredging in Ireland. When Percy died in the early 1940's, he left about £90,000, so all steam cultivation was not unprofitable. On the day the news of his death was told around the district, it was said by some, 'Grundy's dead—steam ploughing is done for'.

XXII

The Steam Plough in World War II

Some steam cultivation was, of course, continued by certain of those owners who had tackle in a workable condition throughout the period of the Second World War 1939–45, though work was only on a very small scale.

A few plough engine fires were relit for a new job in East Anglia. When Warboys, Hunts, aerodrome was laid out hurriedly in 1940, the contractors had to clear some biggish elm and ash trees off quickly from hedgerows and plantations. Somebody hit on the idea of using steam plough engines with their cable haulage to lug the felled trunks into an out-of-the-way corner of the new airfield. I believe at least one engine went from the Whittlesey Steam Ploughing Co., and two owned by Mr How, the auctioneer at Ramsey. It was the last active job How's engines did, and No. 15162 still stands in Upwood village where she was parked after she came home from clearing the way for Lancaster bombers on that aerodrome.

Over at Wittering R.A.F. Station alongside the Great North Road, south of Stamford, the Commanding Officer got into hot water for his initiative in using a pair of plough engines. It was in the early days of the war, and he was appalled at the shocking loss of life and aircraft simply because the old runways were not long enough for the new bombers. He asked for extensions as a matter of urgency. When officialdom in London began to stonewall a bit at his request, he took the law into his own hands. A farmer friend of his nearby had two ploughers, so he called them into immediate commission. They made short work of pulling out trees in a wood just beyond one of the undersized runways, where the standing timber had proved a deathtrap for overshooting bomber crews. There was a bit of a stink about it all, but the danger spot was eliminated and very soon afterwards the runways were properly lengthened as Group Captain Basil E. Embry, D.S.O. and Two Bars A.F.C., had requested.

Perhaps the most spectacular feat of that war, as far as plough engines were concerned, was the cable haulage help they performed on PLUTO, the pipe line under the ocean project. This imaginative scheme was a great success in pumping petrol across the bed of the Channel to supply our advancing armies after their 'D Day' landings in Normandy. The Petroleum Warfare Department were responsible for the project above high-water mark. Two types of pipe were used, Hais cable, which was like heavily armoured 3-inch internal diameter telephone cable, and Hamel steel piping of the same diameter and which could be coiled on 40-foot diameter sea-going drums called Conundrums. It was the intention to tow the loaded 'Conundrums' by sea, and cast them off from their tugs in deep water opposite one of the secret shore pumping stations. The next move was to haul in one of the open pipe ends to the beach. What was wanted, of course, was some mobile form of windlass and somebody thought the steam plough was the answer. In 1942 Messrs Penfolds, steam ploughing contractors of Arundel, received worthwhile hire offers for two sets of their Fowler compounds Nos. 15166/7 and 15220/1. The engines then went over to Harland and Wolf's marine works, where they had their cable drums modified. Special cast iron blocks were fitted to convert the ploughing drums into bollards. This was necessary in order to permit an unlimited length of cable to be wound in. In other words, instead of coiling the cable completely on to the drum, the cable was to have a few winds turned round the drum to give a sufficient grip, and then it was thrown off into a loose coil, just as sailors do when hauling in the mooring ropes of a departing steamer.

Sometimes, the 1,600-ton 'Conuns' with their thirty miles of piping were left a mile off shore. As the plough engines hauled in the pipe ends during the darkness of night, a squad of soldiers stood behind them and coiled the rope as it came off the bollarded-drums. Once the end of the pipe reached the beach, it was then taken in hand by a caterpillar tractor. There were two pipe lines, one from Shanklin (I.O.W.) to Cherbourg, and a later one from Dungeness to a point near Boulogne. The pumping stations were camouflaged as seaside hotels and bungalows. The British and the American armies went over to Normandy on 'D Day', June 6, 1944, and by August 12, oil was flowing through the Shanklin-based line into France. One of the Fowler engines actually went over to France for hauling in the pipeline on that side. She came back with a brass plate recording the foreign mission she had undertaken.

After the war was over, the government became interested in the idea of growing groundnuts on soil of doubtful suitability in Africa. The Overseas Food Corporation had the four PLUTO work Fowlers thoroughly overhauled regardless of cost, and shipped out to break up the virgin land in readiness for the planting of groundnuts. Unfortunately, the whole project was a flop, and the engines were left stranded in West Africa. In 1960, Mr H. L. Drewitt the well-known steam plough owner of Colworth Manor, Chichester, who knew much of the PLUTO story, told me that the four engines were then lying derelict in West Africa. When I tried to get some further information through the West Africa Company, I learnt that there had been some ploughing engines lying about unused in one area, but they were now broken up. Although it was not possible to identify any as previously connected with PLUTO, it was remembered the last job one pair did was to roll down the foundations for some new tennis courts. All that remained at that time as a relic was one plough engine whistle which had been incorporated in the door bell of a Senior Agricultural Engineer's house.

XXIII

Boiler Explosions

A boiler explosion is the most serious and sensational thing that can happen to any steam engine. Fortunately, such events are rare, but there have been occasions in the past where the boilers of steam plough and other traction engines have exploded with great violence. Some idea of the stress placed on boiler plates may be gathered from the fact that with an engine pressed at 160 per square inch there is almost $10\frac{1}{4}$ tons outwards pressure on every square foot of boiler surface, so once a breakage does occur, some weighty and destructive iron parts are blown sky-high in all directions. These mishaps could be caused by a number of factors. The boiler plate may have worn thin due to rust. The plug in the firebox crown might not have been a proper lead-filled plug. Perhaps a group of firebox side plate stays were broken. Or there may have been an old fracture in some part of the boiler or firebox plates.

In 1958, the Rev. R. C. Stebbing of Tacolneston, Norwich, one of our leading authorities on steam plough technicalities, while writing to *Steaming*, gave some interesting details about boiler explosions on plough engines. The following seven instances were known to him:

Fowler	No. 1106 (14 h.p. d.c.)*	5.10.86
	1288 (12 h.p. s.c.)†	1.7.87 (At Martlesham)
	2909 (14 h.p. s.c.)	1.8.13
	3284 (14 h.p. s.c.)	1916 (In Essex)
	3489 (14 h.p. s.c.)	24.4.11
	3616 (16 h.p. s.c.)	15.6.04 (In Catley Park, Linton, Cambs.)
Aveling and Porter	711 (12 h.p. s.c.)	1904

  * d.c. Double crank
  † s.c. Single crank

It was the original practice of Fowlers to make the 3-feet diameter boilers of their single-cylinder engines of $\frac{3}{8}$-inch plate; but after several explosions, they subsequently increased the thickness to $\frac{7}{16}$ inch in order to give a better safety margin. My father used to talk about a ploughing engine boiler which burst in 1879 whilst at work in a field several miles to the east of Sleaford, Lincs. I believe the driver was killed, but the thing my father remembered most clearly was the way in which the explosion turned back the steering wheel on its shaft like the curved handle of an umbrella. Actually, this engine was 1868 Fowler No. 1166.

Fowler No. 3616 which blew up in Catley Park in 1904 belonged to Pamplin Bros of Cherry Hinton. The cultivator had just begun its return journey to the opposite engine when the boiler burst amidships. The front half of the engine was blown forwards some distance, and whilst the driver was seriously injured, the poor water cart man was killed. This accident had an unusual effect on another owner living in Huntingdonshire. Mr William George Thomson lived at 'Providence House' in the little village of Upwood, near Ramsey. He was a deeply religious man who on Sundays used to set out in his pony cart to preach in the various Wesleyan chapels round about. In addition, he owned two sets of Fowler single cylinder ploughing engines which went out on contract work. The sad result of the Catley Park explosion led him to reflect on the possibility that something similar could happen to one of his engines. He decided to take no risks. As soon as the 1904 season was finished, he arranged for each of his engines to be run down to Fowells in St Ives, ten miles away, and have a new boiler fitted. One of the boilers taken out was sent back and converted in his engine workshop close to the house, into a soft water butt. It still stands there. Across the yard, also stands Fowler compound No. 15162 already mentioned in connection with hauling trees off Warboys' aerodrome. A year or so ago, the electric power line supplying the caravan site around her was fixed to her bell-topped chimney, but happily that indignity has now been removed.

It should always be remembered that the boilers of plough engines were subject to two heavy strains. Not only had the internal steam pressure to be contained but there was also the load imposed on the barrels when taking the strain from the underslung rope drums, as the engines did their cable work. The whole of the pull exerted on the cables was taken by the drum stud fixed under the boilers. I do not know whether any engines actually exploded while winding the ropes; but in the case of the Catley engine and another case related to me, it was not so. Once when I was going down in the train to Leighton Buzzard on the way to an August Bank Holiday rally at Woburn Abbey, a chance acquaintance in the compartment said he knew of an instance where some boiler tubes were renewed one Sunday. On the Monday morning, the foreman of the set thought he would have a look to see if any of the tubes were leaking. He waited until the implement had turned to go back before opening the smokebox door, but the instant he did so, the boiler exploded like a bomb. Mr Reginald A. Thompson, of the Upwood family just mentioned, told me recently, that in 1928 he did a season's driving for Mr Briggs of Stamford. When they went into one field he was told by his mates that many years ago a plough engine had blown up there. The driver's short jacket, it seems, had been blown clean off his back and flung high up into the small branches of a tall tree where it remained as a sad reminder for a considerable time afterwards. It had been known there was an old crack in the firebox of the offending engine, but it was hoped it would 'hold'.

Nowadays, although I can see no compulsion in law to have traction engine boilers properly examined at intervals, most owners do take out insurance against possible mishap. As a general practice, a boiler insurance inspector's annual inspection certificate is required by Insurance Brokers, before they accept Third Party insurance cover under the Road Traffic Acts.

XXIV

The Last Chapter of Active Steam
on the Land

One job for which plough engines were much used was river and lake dredging. This work is, in fact, still being done and may well be one of the last bits of contract work ploughing engines ever undertake. Whenever a widish river, reservoir or lake requires dredging for mud, some sort of cable tackle is required, and the steam plough is the obvious choice. Messrs Mornement and Ray, Dredging Contractors of East Harling, Norfolk and of Boston claim that they were first in this field, having begun steam dredging as early as 1899. They specialized in work on the tidal estuary of the Ouse. The engines operated in pairs, just as for land cultivation, hauling the mud scoop to and fro between them as if they were ploughing under water. The King's Lynn Dredging Co. still have steam tackle converted to diesel at work in that area. Bomford and Evershed of Salford Priors have also done a great deal of dredging. They modified their engines by the addition of side drums driven directly from the crankshaft. These drums were used to coil the lighter steel ropes which hauled back the empty scoops, for in dredging, as in mole draining, the loaded pull was one way only. Some of Bomford's steam ploughers were converted to diesel by fitting a diesel engine either in the bunker or over the front of the boiler, after the steam cylinders were removed. These conversions, while removing the need for carting water and coal, spoilt the looks of the engines altogether. As one correspondent said when writing in July 1955 to *Country Life* 'The glamour had worn off'.

Messrs J. B. Carr Ltd, Tettenhall, Wolverhampton began contract steam dredging in 1926 and are still very active in this and similar work. They have eight two-engine sets, mostly workable engines bought at various dispersal sales up and down the country.

A photograph of one of their engines at work pulling up large trees is included in this book. As a matter of interest, Messrs Carr say they consider that of all the Fowler engines they ever possessed, whether for dredging or cultivation, the BB type compounds were the best. A story about an unusual occurrence with a set of their engines is still remembered. Whilst descending a steepish hill, the hind wheel of one engine became detached, and soon rolled ahead at a considerable pace towards a pony approaching with a milkcart. This pony, however, showed great intelligence, because as soon as it saw the great iron wheel bounding downhill it turned about quickly and raced off in the opposite direction. This was an instance where much damage might have been done, but which ended with nothing more than a dent or two on the plough engine which was left standing drunkenly on three wheels.

During the winter of 1959–60, two pairs of Carr's engines dredged the Broadwater Lake in Lord Salisbury's Hatfield Park, Herts. I went out one day by train to look at them, but the tackle was idle at the time. One pair of compounds had 'J. Chivers and Sons' still discernible in faded paint on their bunkers, showing they had formerly been at home on the jam maker's farms at Histon.

The mud scoop in use lifted out 3 to $3\frac{1}{2}$ tons each time. The clearing of the lake, really a dammed and widened section of the river Lee, was a 6-month task and involved taking out approximately 20,000 tons of mud and sediment. This was quite a job for aged ploughing engines to accomplish. Where the old vineyard stood on the north side it was impossible to station an engine because a tall red brick wall came right down to the water's edge. It took three engines to handle the scoop there. The job of the ex-Chivers engines I have mentioned was to stand downstream of the vineyard, and when the 'empty' engine had drawn the scoop down into the water, she pulled it lengthwise down the lake bed to reach the 'dead' length which had to be worked at right angles to the course from where the mud was actually being lifted. The engine to haul out the full scoops was so hidden away in the woods about 100 yards from the lake, that I had a bit of difficulty in finding her. The autumn leaves were just falling from the overhanging oak and sycamore trees, and she looked so strange there in the wood with a littering of coloured leaves on her front tool box.

The last bit of steam ploughing work proper that I saw was in the autumn of 1953 at Ewerby in Lincolnshire. I was spending a few days of late annual leave with my mother down at Barkston. It was early October and I chanced one evening to pop into the Stag Inn for a drink. A stranger was standing in the back yard with a half pint in his hand and we got into casual conversation. I learnt that he lived out Sleaford way and was doing a bit of casual work by way of relieving the Barkston railway crossing keeper for a couple of weeks. The talk turned to agriculture and aiming as it were a bow at a venture, I mentioned steam cultivation of former times. To my surprise, he looked at me and said, 'I can tell you where there is a set working at the moment in Ewerby'. I thought he was pulling my leg, or may be thinking of a pair of tractors at work, but when he said, 'They are a fine pair of engines belonging to Dennis the farmers at Kirton', I then knew he was talking about the right thing. The next morning I caught an early bus to Sleaford, where I changed to the Boston bus, and got off at the Asgarby crossroads, the nearest place the bus goes to Ewerby. I then walked north for a couple of miles. It was a lovely autumn day of Indian summer weather, just right for a day out steam ploughing. Red apples hung by the hundred on cottage garden trees, blackberries by the basketful on the hedgerows and when I turned aside to look into a wayside church, there in the south side porch stood two huge wicker laundry baskets full of the best vegetables and fruit in readiness for the harvest festival. As I walked along the road, I listened for the first sounds of the engines, trying to imagine precisely how I expected they would sound after so many years. I heard nothing till I topped a low ridge looking down on to the parish of Ewerby. Then quite suddenly I heard the distant muffled, fast rumble of the engine that was pulling—the whirring sound that comes from compounds at work. This sound made me turn to the north-west and sure enough there under a thin mist of smoke were the unmistakable black outlines of two plough engines, one on either side of a field.

In the village there was a combined post office and general shop. The lady who kept it was a chatty soul and when she heard I had come to look at the cultivating engines, she got really animated. 'Do you know,' she said, 'last week we even had two sets at work—side by side in two adjoining fields.' I asked to whom the other set belonged and she told me they were actually owned locally by farmer Herbert Thorlby, over at Eveden. He was, it seemed, a bit of a steam man, with two sets, but had just finished the season's work. After a glass of beer and a sandwich in the pub, I pushed off across the fields to Dennis's engines.

I spent a very happy afternoon with them. It was an experience made more enjoyable and unforgettable simply because I had thought I should never again see plough engines at work. They were ploughing mustard stubble—hard dry ground it was too. The driver of the engine on the right-hand side of the field told me it was several years since the engines had done any work. They were new in 1928 and had never really done a great deal of work, but for some reason or other the boss had decided to send them from Kirton near Boston to do the autumn ploughing on his land at Ewerby. I couldn't help noticing how brisk the compound was, after my early recollections of single-cylinder Fowlers. With its short connecting rods and balanced crankshaft the compound sprang into life at the first nudge on the regulator. She was extremely lively, and whereas the single engine is slow to start the compound slung herself into action. Once the plough was on its way back and full steam turned on, the engine raced and raced till the motionwork and drum winding gear just hummed. It was exhilarating to be on her. The water cart arrived behind a Fordson tractor driven by a tall young man wearing a flattened cloth cap. He seemed to regard the steamers with a sort of benevolent amusement as though they were museum pieces brought out for a day or so's fun. His talk soon turned to some remarkable aerobatics he had seen recently at an R.A.F. display farther north in Lincolnshire. I walked over to the other engine and while there the ploughman invited me to 'have a ride back'. It was a jerky run and the plough had to be held hard on a straight course and, besides, one of the farm chaps was riding as an extra man to help hold the rear end down into its work.

I finished off the afternoon with the driver I had first visited. He was a roundish man of about 60, who said he had, 'come back to cultivating after he finished soldiering in 1919'. He was cheerful and complacent, satisfied to concentrate on his engine and while 'pulling over' never took his eye off the approaching plough. He told me he could remember when they had been so busy with Dennis's engines that they had still been in the field in January and even February. Sometimes the land then had been so wet and sticky, that it had taken all day to get the far side engine across the field and out of the gateway on to the hard road. When the Boston

Peace Celebrations were held in 1946, he had driven one of the Dennis Plough engines in the victory parade, in which they represented 'Old Time Stuff'. Incidentally, the two engines concerned now stand in Ward and Dale's old field at Sleaford, still displaying the little commemorative plaques they received that day. I liked this driver's talk. What a pleasure it was to have the privilege of hearing from an old and experienced man, the tales of steam ploughing—the real stuff! I left him saying, as he had done several times that afternoon whenever the engine jerked at the rough pulling, 'Rum old land round here!' This job proved to be the swan song of the Dennis engines. Next year all their plough sets were sold by auction at Kirton. So ended a long ownership of steam ploughing engines by these important Lincolnshire farmers.

The story of steam ploughing would be the poorer without some mention of John Patten of Little Hadham, Herts. He was a farmer who owned several sets of engines, one of which was out at work when he died in March 1960, aged 80. He was a bachelor and an 'old steam man'. It is said he bought his first set of tackle following an argument with his usual cultivating contractor who had kept him waiting on one occasion many years ago. 'All right,' he said, 'I'll buy a pair of engines and have my work done when and how I want it in future.' By 1918 he was the proud owner of some eleven sets, and had made quite a name for himself as a steam ploughing contractor in the Bishop's Stortford area. As the years went by and tractors gradually usurped the steam work, his sets dwindled, but all through the 1950's he could say that at least two of his sets had done some work each season. Steam traction engine enthusiasts knew he still kept steam ploughing going and often parties were made up to visit his engines at work in the fields.

Apparently he had arranged in his will that his steam tackle should be sold. The eight remaining engines, a derelict hulk or two and implements were auctioned on June 28, 1960, at his former home, Hall Farm, Little Hadham. I went down to the sale. The house is quite close to the flint-faced parish church. In the churchyard I found a bricklayer and his mate mixing mortar for some minor repairs on the church, and I asked them about John Patten. They told me he was well known around the district as a steam plough contractor and one who had kept the old engines at work for old time's sake. He liked to see steam ploughing engines in the fields. He had also been a keen churchman and a benefactor of the little church just outside his garden wall. He looked after his workmen too; I was told that every Christmas until 1939, when the war and food rationing upset things, he had two bullocks killed to give a present of beef to those he employed. 'Once,' the bricklayer said, 'he asked a man whether he had got his beef all right. "Well! yes, thanks, I have," the chap replied, "but it was a bit bony this year".' John Patten took an exception to that remark and the man concerned was not on the next year's list.

From the church I went over to the eight engines lined up for sale below the farm outbuildings. The long thin strip of land upon which they stood had been the usual steam plough parking place. Between the engines and a large field of waving barley just in ear, was a littered miscellaneous collection of rusted and discarded steam implements, harrows, rollers, mole drainers, cultivators and the frames of ploughs, all now reduced to nothing more than scrap. The eight plough engines, six AA and two BB types, were each in steam. As their smoke drifted away over the barley, it gave life and movement to the scene of the sale. The two BB engines had been driven in from work after John Patten's death. They looked splendid. They were clean, their brass and steelwork well polished and the coils of their recently used ropes shone like silver. The one remaining steam plough crew was there to see the end. Although there were but a couple of hours before the engines would have new owners, this or that chap, spanner in hand, still tinkered with the motionwork. That, I feel, is what might be called being devoted to one's job. The six men then posed for a photograph.

It is not very easy for an ordinary chap like myself to distinguish between an AA and a BB type engine. The AA is a little bigger and rated at 18 n.h.p. instead of the 16 n.h.p. of the BB. The distinguishing features of the AA are a slightly chubbier smokebox, double slide bars, and a tail rod on the low pressure piston. The BB engines were built without tailrods, but the later BB1 class had them. The Patten AA engines had handbrakes. All the engines were fitted with outsize drums in order to cope with the wide fields of Hertfordshire. Two engines had empty drums, but the other six carried 800-yard ropes. In general appearance, the AA engine seems to fall a little behind the BB which has a superb outline. The BB job seems perfectly balanced and beautiful. There is little doubt these particular Fowler plough engines were the best designed of all traction engines. They combined elegance with mechanical perfection and looked and sounded right in every way. I do not know the name of the man to whom should be given the credit for this great and unsurpassed piece of design work at Fowlers.

I had a chat with the gang foreman. He told me how, years ago, when a young man he worked for a Bedfordshire ploughing contractor and sometimes new sets of tackle would be driven up by road from Leeds instead of being sent by rail. As the men came south with the engines, it was usual to lay up overnight in the market-places of country towns on the Great North Road. I mentioned that I had at one time heard there was an agreement that if ploughing engines were used in a field where there were electricity supply posts or pylons, a sum of 5/- for each post could be claimed from the Electricity Board. He confirmed that it was his understanding this still applied to compensate for the inconvenience and delay in unhitching the ropes in order to pass these obstructions. The final wage rates at Little Hadham were £8 5s. a week, with a 2/3d an acre bonus shared by the gang. When mole draining three yards apart, they had been paid 2/5d an acre. His final remark was that it seemed a pity the engines were being sold for the benefit of Cats and Dogs Homes, for such it seemed were the dispositions of the will.

The sale attracted a few hundred people, including half a dozen ladies. It was a gathering of many old hands in the steam trade though there was also a good sprinkling of younger enthusiasts. You could see old acquaintances meet on the roadway in front of the engines, shake hands and then turn to look or point at the long line of veteran ploughers. On a day such as this, one could almost guess their conversation—'Sad to see the old engines go', or 'don't suppose anybody will buy them for work, but you never know'. Those who had a mind to buy could be seen peering into fireboxes looking for such faults as patches or burnt down stay heads. Some men were 'opening up' a bit just to see how the motionwork turned over.

The youngish auctioneer stood on each engine as he took the bids. In accordance with instructions, they were sold, oddly, without implements, but at the second engine of one pair he did say with some feeling, 'Don't part her from her pal, they have worked together for many years'. As will be seen from the following lists, all went away as pairs.

Type	No.	Name	Price	Sold to
AA	15362	Lion	£215	H. Hardwicke, Elwell
,,	15363	Tiger	£170	,,
,,	15358	Darby	£230	,,
,,	15359	Joan	£200	,,
,,	15364	Windsor	£255	,,
,,	15365	Sandringham	£205	,,
BB	14383	Prince	£310	J. Branch, Cambridge
,,	14384	Princess	£320	,,

Of the four brown-painted living vans on iron wheels, the best fetched £50. The cultivator belonging to the BB set made £420, because these implements were at that time in demand for use behind heavy caterpillar tractors. I bid £49 for the van which made £50. I would have liked an engine but the prices were higher than I had expected. Another chap and I tried to get together and buy back 'Tiger' from Mr Hardwicke, but the price was beyond our combined purses.

As it was perhaps the exception rather than the rule to name ploughing engines, it was surprising to find the whole of this stud were 'Namers' as the train spotters say. Maybe it speaks of the interest and affection John and his crews displayed towards their engines. A year or so later when I went past the farm in my car, the old steam plough parking place had been incorporated indistinguishably into the field. A crop of young green corn stood uniformly over the site, and a stranger today would never guess that anything of interest in iron and steel had ever stood on the hillside there. All trace of John Patten's steam ploughs had gone for ever.

XXV

What was Wrong with Steam Ploughing?

I have already mentioned some of the shortcomings associated with steam ploughing. However, when I was talking recently to Mr Douglas R. Bomford about this book and steam ploughing in general, he suggested I should deal at greater length with this side of the story. He feels that if the manufacturers had been a bit more imaginative and progressive in outlook and ideas, say from 1900 onwards, ploughing by steam might have remained an economical proposition for at least another decade. Why was it, therefore, that the big compound plough engine was replaced so quickly by the motor tractor?

One reason, of course, is that after John Fowler's death in 1864 there was no real improvement in the basic design of cable type plough engines. It is admitted that compounding had been accepted, but the fact still remains that there had been no really important or revolutionary advance. Had John Fowler lived longer it might have been much different. The plough engine, like the railway locomotive, never utilized anything better than the locomotive type boiler used by George Stephenson. These boilers were hardly the best choice for meeting the alternating steam requirements of a plougher. For one thing, they needed firing very carefully, otherwise cold air was drawn into the fireboxes with the result that tubes and stays leaked. There was also the catchy business of keeping the right water levels during 'Head up' or 'Head down' working. Some enlightening experience was gained with a more up-to-date steam generator when the Rhino and Bomford-built lightweight steamers with Sentinel superheated upright boilers were tested in the fields at Pitchill. It was particularly noticeable, I am told, how easily and carelessly steam could be maintained at 260 lb. on these progressive and interesting engines. Another feature which showed up on the Bomford/Sentinel cable

194

engines was the way the ropes coiled perfectly on the vertical drums without any coiling gear. If the big plough engine had, in fact, ever lost its locomotive type boiler as a result of modernization, it rather looks as if this would have been taken as an opportunity to hang the rope drums vertically on any new design. As a rule the Fowler coiling gear worked all right, but when the cultivation was along a hillside, unless the headland upon which the engine stood was at the same inclination as the rest of the field, there was a tendency at times for a coil to drop out of place. When the rope did go 'haywire', not only had the driver to stop his engine and get down to crowbar the erring coil back into position, but there was also the consequent loss of time for the whole gang while the job was stopped.

As any steam engine with a really efficient valve mechanism ought not to need road gears, the question arises whether or no the heavy gearing could have been eliminated on ploughers? Sentinel patent type engines were well designed and exceptionally good performers, and those fitted to the Bomford cable machines neither had, nor needed, any road gears. It might have made an amazing difference to performance if some of the big engines had been equipped with an improved valve gear incorporating long travel valves and constant lead. Any weight saved in the removal of gear trains or coiling mechanisms would have been invaluable in producing a lighter engine altogether. What was really wanted, of course, was an engine with the power of a Fowler BB 16 n.h.p., at only half the weight. An ideal engine of this kind would have got about the fields more easily and at the same time been able to work a longer season. Just think, too, what a satisfactory condenser might have saved in water haulage, boiler priming or boiler washing. Nor must we overlook what the effect of a condenser could have been in avoiding damage to fireboxes because of overheating in the lower levels by the build-up of mud from the raw and often dirty water so constantly fed into the boilers under ordinary conditions. Fuel bills, too, would have been cut.

The plough was a bad implement. Its weight was mainly on the two main wheels, instead of on the ploughshares. We know already about the need to have one or two extra men riding in order to provide end weight. Owing to its length the plough was awkward to handle on the road, and it was sometimes difficult to get it through the wooden-posted gateways of the farmers' fields.

However, in spite of its unwieldy length, the five or six furrows fixed at either end were actually set too closely together. This caused 'choking' whenever there was much top rubbish on the land. The additional man riding on the plough in the illustration of Henry Warner's tackle has a spud in his hand as he tries to push the stubble litter clear. Another point of interest about the ploughs was that they did not always absorb all the available power of the engines. An AA or BB compound could have handled a seven- or eight-furrow plough quite comfortably under most field conditions.

One of the drawbacks attached to all cable cultivation was the leaving of wide unworked headlands when using the plough. The strips upon which the engines stood, plus the length of the plough while at rest, were up to 30 feet wide. In order to plough these headlands before leaving the job, it was necessary to drive one engine half-way round the field to get the pair into the opposite position. If the land was at all wet, this could be more than trying. Of course, whether ploughing or cultivating, there were usually tracks or wheel marks left by the engine which after the finish of work had to 'come over' from the far side of the field. I once heard an old farmhand say how he remembered spending days on end as a young man with a two-horse team, 'Ploughing up cultivator engine wheelings'. After the disappearance of the horse from our farms in the 1930's, this finishing off, as well as the water carting, was done by a tractor.

Unfortunately, the mechanical efficiency of the steam plough engine did not compare very well with the motor tractor. A compound plough engine had an efficiency of about 6%, i.e. the amount of energy in the fuel it burnt actually turned into useful work at the implement. On the other hand, the diesel tractor can claim an efficiency figure of 25%–30%. This means roughly that if a farmer today steam-ploughed ten acres, his bill for 15 cwt. of coal consumed by the pair of engines would not be less than £8. A diesel caterpillar tractor, handled by one man would do the same job in roughly the same time on twelve gallons of diesel oil, costing about 18/–.

So far as the engines were concerned, however, I think any serious attempt to re-design them would probably have spoilt their looks altogether. The beautiful balance of round shapes in boiler, smokebox, chimney, cylinders, rope drum, flywheel and road wheels would have gone. The big compound steam plough engine at work was a glorious thing for anyone to see. I for one am happier to have known her just as she was, old fashioned, heavy and cumbersome and uneconomic, but possessing that undefinable beauty that is the heritage of any 'live' machine.

XXVI

Preservation Days

Although the death of John Patten may have marked, more or less, the end of contract steam cultivation, it did not mean all steam ploughing had finished. At the present time (1963) there are probably some thirty sets in workable order in Great Britain. These tackles are in the hands of various kindly-minded owners and where they have farm land, the engines are used now and again for fun or work as the case may be. Quite a few plough engines have been preserved oddly enough by people who could afford only one engine, and this has broken up a few pairs of engines which otherwise could have been used on the land.

For those who have never seen a cable type ploughing engine, the summer rallies now held by various clubs up and down the country, offer the best opportunities for seeing one. If possible, the organizers like to get at least one plougher on their list. What splendid engines they look, too. Other engines may be elegant, stylish, nippy, or just useful, but I always like to see a big Fowler compound roll forward off its stand and set forth majestically along the rows of spectators. They are peerless things in steam.

At the Appleford 1960 rally organized by the National Traction Engine Club, some ploughing and cultivating were carried out as a special demonstration feature. We had a Fowler BB pairing a McLaren compound, and another BB and a Fowler single mated up as a second pair. The first two engines worked a balanced plough and the other two had a cultivator. The exhibition was given in a field adjoining the rally site on Arthur Napper's farm and it was most encouraging to see how much interest was aroused among the spectators. Old steamploughmen, long since retired, waxed enthusiastic as the big engines began to take up the strain and 'chunter' at the pull. Many who were seeing the job done for the first time in their lives asked me, as the commentator, how long it was since it went out of fashion to do ploughing that way.

198

Similar shows have been staged at other rallies and no doubt for as long as the engine boilers can be steamed, there will be further displays.

Mr H. Jackson, a farmer of Toppesfield, Essex, has owned a set of AA engines since 1918. He still likes them and believes in them. To his mind the steam engine does his autumn cultivation better than anything else, so after each harvest, he steams up as a matter of normal routine. Messrs Boughton and Sons, of Amersham, Common, Bucks. with steam plough associations in the family (the late Mr T. Boughton was a grand old steam man), usually take a pair of engines out into the field in September or October. The names I have mentioned, of course, represent but a few of those dedicated people, who continue to keep the memory of steam ploughs green by showing working engines to all who care to go and see them. It just goes to show the extraordinary appeal and fascination of the steam cable cultivating engine.

In the year 1918, when the steam plough was at the height of its popularity and usefulness, Britain, too, was at the peak of her role as a world power. Steam power made Britain, and unfortunately the two have since declined together. However, I hope it will be long ere the last dish-spoked flywheel ceases to turn on a plougher or the last live steam tale is told.

APPENDIX

PRINCIPAL IDEAS AND ACHIEVEMENTS ASSOCIATED WITH STEAM CULTIVATION: 1619–1860

Year	Inventor	Patent No.	Machinery
1619	David Ramsey and Thomas Wildgoose.	6	Undefined engines for ploughing without horses or oxen.
1769	Francis Moore, Cheapside, London. (Linnen draper and Warehouseman)	921 & 933	Claimed coming 'Fire Engine' would displace horses at the plough. He sold his own horses and advised his farmer friends to do likewise.
1784	James Watt, Birmingham.	1432	A steam carriage for moving persons, goods or other matters from place to place.
1810	Major Pratt, Spencer Street, St George in the East, London	3309	Cultivator with rotating tines, as well as a two-way plough for steam or other suitable power.
1813	Richard Trevithick, Camborne, Cornwall.	—	Rotary type cultivator for haulage behind a steam engine.
1816	Joseph Reynolds, Retley, Wellington, Salop.	3973	A steam carriage to pull implements.
1822	Mr Roberts.	—	Digging machine equipped with rotating tines, but probably intended for horses.
1829	Adam Scott.	—	Invented mole draining implement.
1829	Henry Handley, Culverthorpe Hall, Lincs.	—	Offered a prize of a hundred guineas for the invention of a satisfactory steam plough.
1832	Joseph Saxton.	6351	System of ropes and pulleys for the better application of power to agricultural implements.

Ideas and Achievements associated with Steam Cultivation

Year	Inventor	Patent No.	Machinery
1832	John Heathcoat, M.P. and Josiah Parkes	6267	Designed, built and demonstrated a steam plough with power unit on tracks and plough hauled by a flexible iron-plated belt. This was the first practical steam plough: actually introduced 1833.
1834	Marquis of Tweeddale, Yester, East Lothian.	—	Asked the Highland & Agricultural Society to encourage steam plough inventors by offering a money prize.
1837	John Upton.	7458	A steam land carriage having a rotary type engine. Vehicle designed to haul agricultural implements.
1837	Highland & Agricultural Society.	—	Offered a premium of £500 for a successful system of steam cultivation. This offer was withdrawn in 1843 without any award having been made.
1839	Alexander MacRae, Demarara, British Guiana.	8329	Single engine system specially designed for engine and opposite anchor to be carried in punts.
1839 & 1840	Henry Pinkus, Panton Square, Coventry Street, London.	8207 & 8644	Two-way plough moved by vacuum operated engine. Vacuum to be supplied from central steam powerhouse through underground pipes.
1842	W. Blurton, Staffordshire.	—	Experimented with a steam engine which pushed its plough.
1843	A. Atkins, Chepstow.	—	Rotary digger for cable haulage—power to be obtained from a stationary steam engine.
1846	T. Bonser and E. W. W. W. Pettitt, Merton, Surrey.	11297	Rotary cultivators with prongs and curved blades for horse haulage, pending invention of suitable steam engine.

Year	Inventor	Patent No.	Machinery
1846	John Tulloh Osborn, Demarara, British Guiana, and Marquis of Tweeddale, Yester, East Lothian.	11304	Two-engine steam ploughing tackle with two vertical cable drums alongside the boiler of each engine.
1847	Pierre Philippe Celestin Barret, Paris.	—	Patented a steam grubber.
1847	Sir John Scott Lillie.	11907	Implement with rotating tines intended to be pulled about by a steam engine.
1849	James Usher, Edinburgh.	12710	Combined a mobile steam engine with a rear-mounted revolving digger.
1850	Lord Willoughby d'Eresby, Grimsthorpe Park, Nr. Bourne, Lincs.	—	Experimented with both single and double engine ploughing sets. The engines had underslung drums which coiled link chains.
1850	Mr Hannam, Burcote, Abingdon, Berks.	—	Designed the first roundabout style of tackle, which was built by Messrs Barrett of Reading. Portable steam engine and iron wire rope.
1850	Major Amos Tyrell, U.S.A.	—	A combined steam engine and wheel plough.
1851	Chandos Wren Hoskyns.	1899	Propounded an idea for a steam land rasp which he patented in 1853.
1852	John Bethell.	949	Constructed a rotary digger which had a steam engine mounted on the same frame.
1853	William Smith, Woolstone, Bucks.	2121	Bought an engine and windlass and commissioned a heavy cultivator from Messrs Howard of Bedford.

Year	Inventor	Patent No.	Machinery or Association
1853	John H. Johnson, 47 Lincoln's Inn Fields, London and Robert Romaine, Peterborough, Canada.	1151 35 (1855)	Horse-drawn cultivator on rotary action principles with steam engine on frame to power the spiked digging drum. Vertical boiler, two cylinders fixed 'Shay' style at one side.
1854	John Fowler, Jnr., (The Younger), Temple Gate, Bristol (Agricultural Implement Maker).	1570	Exhibited a steam mole draining machine at the Royal Agricultural Show, Lincoln.
1855	David and Thomas Fisken, Engineers, Stockton-on-Tees.	1629	Double-ended plough with system of bell cranks and levers for raising and lowering frame at ends of field.
1855	P. A. Halkett, Wyndham Club, 80 Chancery Lane, London.	2224	Constructed a set of tackle on the 'Halkett Guideway' system. Two steam engines running over iron rails laid across fields permanently.
1856	John Fowler (Jnr).		Demonstrated successfully at Nacton in Essex, using portable steam engine with balance plough moved to and fro by cable on roundabout layout.
1856	Mr Holcroft.		Patent for Screw propelled steam plough. No record of actual construction. (Holcroft was impressed with screw propulsion of steamships.)
1857	Thomas Rickett, Buckingham Iron Works, Buckingham.		Designed and built a steam-driven rotary digger.
1857	J. Bethell, 8 Parliament St, Westminster.	2691	'Boydell' type wheeled digger or forker.

Year	Inventor	Patent No.	Machinery or Association
1857	W. H. Nash, Cubit Town, Isle of Dogs, London.	—	Traction type digger powered by two cylinder steam engine. It had hydraulic rams for lowering and raising the digging tines.
1857	C. Hall and T. Charleton Knavestock and Brentwood, Essex.	—	Patent for direct traction steam plough equipped with centrally-placed hydraulic jack upon which engine was turned at ends of fields. No record engine was built.
1859	Chandler and Oliver, Bow, London.	—	Fitted drums to extended axles of portable engine for round-about style ploughing. Tackle made by Robey & Co., Lincoln.
1860	Henry Grafton.	—	Designed a system in which two engines, on caterpillar tracks, were connected by an overhead girder from which implements were hung, as with Halkett's tackle.

Note on the horse power of engines

The use of n.h.p., meaning nominal horse power, for classifying the power of engines should be accepted only as a basis for comparison of the relative power of the various types of engines. Nominal horse power, as given by the engine manufacturers, has no uniform relationship with the true or indicated horse power, which could be between five and ten times as great.

Index

Acreage payments, 44, 59, 111, 118, 192
Aerodrome work, 180
Aeroplanes, 137, 165, 176, 180
Agricultural, General Engineers Ltd., 68
Allen, John, 110, 112, 115, 140–5
'Allens Activities', 110
American Civil War, 52
Anchor cart, 38
Ant heaps, 151
Army recruitment, 141, 143
Arnold, Matthew, 109
Atkins, T., 37
Australia, 81
Austria, 148, 151
Aveling & Porter, 58, 68–71, 150, 172

Barkston, 73, 78, 102, 117, 126, 176, 188
Barrett & Exhall, 32
Barry, J., 21
Barrows & Stewart, 58
Bateman, Dick, 165
Beeby Bros., 113, 168
Belt work, 36, 62, 73, 74, 124, 153
Berners, Lord, 57
Bethell, J., 32
Bible, The, 50
Big Hole, Kimberley, 149
Blacksmiths, 19, 108, 111, 114
Bog oak, 175
Boiler (Multitubular), 83
Bomford, Bros., 53, 60, 106, 156, 157, 158, 186, 194
Boiler washing out, 133
Borsig, Messrs., 162
Boughton, Tom, 199
Boydell, J., 51
Brakes, 72, 86, 87, 115, 133, 173
Bridges, 147, 167, 168
British Empire Exhibition, 1924–5, 172
Burrell & Co., C., 58, 68, 69, 75, 76, 80, 94

Cadwell, C., 92
Canada, 33, 91, 95, 125, 153

Carr, J. B., Ltd., 186, 187
Case, Jerome Increase, 96, 120
Caterpillar Tracks, 21
Chain Drive, 40, 72, 75
Chandler & Oliver, 60
Clark, Ronald H., 125
Clayton & Shuttleworth, 36, 39, 40, 52, 58, 121
Clip Drums, 40, 49
Clough, Arthur, 109
Coal consumption, 113, 143, 158, 170, 171, 195, 196
Coleman & Morton, 52, 58
Collings, J., 124
Compound engines, 76, 78, 79, 113, 114, 122, 155, 167, 172, 189, 196
Condensing of steam, 76, 195
Contractors, 61, 62, 64, 66, 72, 108–19, 172, 186, 198
Cooper, T., 91–3
Costs and charges, 30, 57, 58, 60, 61, 62, 63, 72, 73, 90, 121–3, 140, 149, 170, 172
Cotton, 151
Couch grass, 102
County Food Production Committees, 137
Country Life, 186
Cowley, 108–14, 142
Cuba, 102, 153
Cultivator, 18, 60, 61, 100, 102–3, 104, 106, 193

Daily Chronicle, The, 68
Damage to roads, 168
Darby, T., 89–91, 122
Davey, Paxman & Co., 90
Deep Ploughing, 148, 151, 175
Demarara, 24, 25
Dennis & Sons, 171, 188–90
Diesel engine, 83, 160, 186
Diesel, Rudolph, 160
Differential Gear, 72, 115
Digger, 33, 88–94
Discer implement, 153
Dorchester, 110, 112, 114, 115
Dredging, 179, 186, 187
Drewitt, H. L., 182
Dynamite cartridges, 55

Eddington, W. & S., 89
Eddison Steam Ploughing Co., 114
Eddisons, The, 76, 108–10, 112, 114, 115, 146
Egypt, 52, 54, 70, 81, 83, 93, 106
Electric Ploughs, 164
Electricity Pylons, 192
Embry, Group Captain Basil, 180
English Social History, 109
Ewerby, 188
Eyth, Max, 53, 54, 55, 159, 164

Fallows, 64, 104, 105
Fawkes, J., 95
Feed water heater, 72, 122
Fisken, W., 37, 50, 51, 72, 80, 81
Flywheels, 47, 77, 81, 85, 111, 130
Fodens, Messrs., 157
Food Production Dept., 142
Food Rationing, 143
Ford, Henry, 99
Foremen, 61, 107, 110, 111, 117, 118, 127, 139, 175, 192
Fowler, John, 34–42, 49, 53, 54, 146, 164, 194
Fowler & Co., John, 39, 45, 48, 57, 58, 69, 76–80, 91, 102, 103, 120, 142, 146, 159, 163, 172
France, 69, 70, 80, 102, 148, 165, 182
Franco–Prussian War, 62, 78
Friction Drive, 77, 114, 115
Fusible Plug, 85, 163, 183

Gallipoli, 165
Gang Plough, 24, 97, 155
Garrett & Co., R., 53, 68, 121, 124, 154
Gear mechanisms, 67, 75, 77, 86, 130, 195
Gears, ringing note, 128
Germany, 53, 54, 55, 75, 102, 148, 159
G.N. Rly. Co., 72, 73
Gooch, Daniel, 27
Grantham, 22, 28, 127
Great Exhibition of 1851, 30
Grieg, D., 36, 51, 62
Grimsthorpe, 28, 29, 38
Groundnuts Scheme, 182
Grundy, Percy, 144, 169, 173, 179
G.W. Rly., 27, 68
Guideway System, 51

Halkett, Lieut. P. A., 51
Hanley, H., 22

Hannam, 31
Hariali grass, 151
Harrows, 51, 64, 72, 100, 117
Hawaiian Islands, 102
Hayes & Co., 58
Heathcoat, J., 20, 23
Hengist and Horsa, 71
Heucke, A., 163
Highland & Agricultural Soc. of Scotland, 22, 50
Hilditch, J., 51, 52
Hornsby, R., 28, 58, 72
Horses, 20, 25, 26, 33, 39, 44, 57, 61, 62, 64, 104, 109, 117, 166, 187
Horse-steered engines, 39, 60
Horsefield & Co., 58
Howard, J. & F., 32, 48, 57, 58, 65, 66, 67, 93, 122
Howe, W., 84
Hydraulic Jack, 90

India, 151, 160
I/C Tractors, 25, 98, 102, 114, 120–4, 137, 145, 146, 150, 154, 158, 170, 172, 174, 189, 196
Illinois State Fair, 95
Injector, 84, 134, 136
Italy, 102, 148

Japan, 151

Kantenden soil, 151
Kemna & Co., 159
Kenelly, Susannah, 166
Kimbell, E. E., 70, 144
King Alfred's Hill, 60
Kitchener & Co., T. B., 174
Kitson & Hewitson, 39, 40, 58
Knifer implement, 151, 152
Kruger, President, 149

Lang's Lay, 107
Last Fowler Plough Engines, 173
Lavergne, L., 33
Licences, 173
Living Vans, 59, 108, 114, 129, 134, 143, 150, 177, 193
Lloyd George (The Premier), 142
Lochar Moss, 22
Locomotive Act, 1865, 117, 143
Locomotive Act, 1898, 167, 169
Locomotive Act, 1878, 169
L.N.E. Rly., 66, 157, 161

McLaren, J. & H., 80, 81, 90, 91, 122, 123, 172
MacRae, A., 24
Manilla ropes, 72
Manns' Patent Steam Cart & Wagon Co., 122
Marshall Sons & Co., Ltd., 120, 124
Mechanics Magazine, The, 66
Mechi, Jon, 33
Melksham, 34
Memorial to John Fowler, 41, 164
Mexico, 153
Midland Engine Works, 81
Mishaps, 48, 87, 136, 138, 150, 166, 167, 168, 183–5, 187
Mole drainer, 35, 100, 147, 171
Moore, F., 17
Morris, J., 88
Motorists, 141, 167
Murdock, W., 19
Mussolini, Benito, 148

Nacton, 37, 38, 42
Napper, Arthur, 198
National Traction Engine Club, 91, 164
Navigational aid, 166
Newcomen, T., 17, 18
N.H.P. (nominal horse power), 204

Oil burning, 76
'Old Abe', 97
Osborn, J. T., 25
Oxen, 20, 43, 44, 61, 149
Oxfordshire S. P. Co., 110, 112, 174

Pacific Rural Press, The, 95
Paintwork, 173
Pamplin, Bros., 141, 184
Parkes, J., 21
Patten, John, 190–3
Pepper, Alf, 80
Peru, 102, 152, 160
Philippine Islands, 102
Piccadilly, 142
Pigs, 62, 73
Pinkus, H., 25
Plough, anti-balance, 100, 101, 195
Plough, disc, 102
Plough (Turn-rest), 37, 104
Plough Monday, 43
Pluto pipeline, 181, 182
Pontine Marshes, 149

Portable Rly., 29
Portable engine, 31, 37, 45, 60, 89, 96
Pratt, Major, 18, 19
Proctor, F., 93
Prosecutions, 69, 141, 166, 169, 174

Rallies, 198
Rampant Horse of Kent, 71
Ramsey, D., 17
Ransomes (Ipswich), 25, 37, 39, 58
R.A.S.E., 45, 47, 56, 58, 62, 121
Rate of work, 22, 30, 60, 90, 104, 122, 123, 156, 187
Ravensthorpe Co., 51, 72, 80
Rheinmetall, Messrs., 163
'Rinoceros' engine, 156, 194
Rice, 151
Richardson & Darley, 58
Roads Act, 1894
Roads, H., 178
Robey & Co., 52, 58, 60
Romaine, R., 33
Roo Dee, 40, 47, 62
Rope anchors, 58
Rope coilers, 41, 55, 80, 195
Rope porters, 75
Ropes, Iron wire, 82
Ropes, steel wire, 39, 60, 70, 79, 88, 106, 107
Rose, Joe, 118
Rotary cultivator, 18, 19, 27, 32
"Roundabout" cultivation, 31, 32, 39, 58, 60, 66, 74, 76
Royalty Co., 40
Ruston & Co., 58

Safety valve, 84, 112
Savage, F., 69, 74, 90
Savery, T., 17, 18
Savory & Son, 52, 58, 60
Schmidt, W., 160
Sentinel engines, 154, 156, 157, 194
Single cylinder engines, 77, 78, 79, 113, 135, 189
Slant Shafts, 77
Sleaford, 102, 116, 118
Smith, W., 32, 36, 57, 58, 66
Smith (of Coven), 58
Smoke, 69, 145, 166
Smoke-boxes, 112
South Africa, 70, 93, 102, 149, 150, 156
South Pacific, 152

Speed of work, 63, 80, 101, 104
Splicing wire ropes, 107
Standish, P., 95
Stebbing, R. C., Rev., 109, 183
Steam Cultivation Development Association, 140–7, 167–70, 178
Steam Plough Inn, 31
Steam Plow Contests at Winnipeg, 98
Steam rollers, 68, 112, 115
Stephenson, George, 39, 50, 84, 194
Stephenson, Robert, 50
Stileman's Works, 91
Straw burners, 148
"Suffolk Punch", 154
Sugar cane, 151, 152
Summerscales engine, 154
"Superba" engine, 149
Superheater, 122, 123, 155, 160–3
Sutherland, Duke of, 54
Syston, 72, 73

Tandem engine, 124
Tanks (British armoured), 21
Tasker & Co., 58, 121
Teeterboard, 98
Thompson Road Steamer, 88
Thorold, Sir John, 72
Three-wheeled engines, 66, 96, 154
Times, The, 31, 38
Trade depression, 63, 108, 145, 172
Training, 59, 144
Tramway Locomotive, 19
Trenches (Military), 165
Trevelyan, G. M., 109
Trevithick, R., 19, 20
Tube expanders, 144
Tuxford & Sons, 71, 72, 74
Tweeddale, Marquis of, 26

Upton, J., 24
Upwood, 180, 184, 185
U.S.A., 76, 95–9, 120, 125, 153
Usher, J., 27

Vacuum Power, 25
Valve gear, 77, 84, 85, 93, 122, 130, 154, 156, 195

Wadkin, Joe, 128, 129, 138
Wages, 30, 74, 118, 142, 144, 192
Wallis & Steevens, 121, 123
Ward & Dale Ltd., 115–19, 127, 129, 133, 175–8
Water carting, 57, 58, 59, 77, 113, 132, 195
Water lifter, 86
Water Tube Boiler, 24, 66, 67, 93
Watt, J., 17, 18, 19
Weston Park, 91
Wild animals, 97, 151
Williams, J., 40
William The Conqueror, 71
Willowby, d'Eresby, Lord, 27, 28, 30, 38
Windlass, 35, 36, 37
Wittering, 180
Wolf, R., 163, 164
Worby, W., 36, 37
Worcester, Marquis of, 17
Worms, 58
Wright, Philip, Rev., 164

Yarrow, A., 51, 52
Yorkshire Patent Engine Co., 156